ELECTRONICS

ELECTRONICS

A systems approach to digital electronics

G. E. Foxcroft O.B.E.
Formerly Senior Science Master, Rugby School

J. L. Lewis O.B.E.
Formerly Senior Science Master, Malvern College

M. K. Summers
Department of Educational Studies, Oxford University

Longman

LONGMAN GROUP UK LIMITED,
Longman House, Burnt Mill, Harlow,
Essex CM20 2JE, England
and Associated Companies throughout the world.

First published 1986

Set in 10/12 point Times
Printed in Great Britain
by Scotprint Limited, Musselburgh, Scotland

Contents

	Preface	vii
Chapter 1	Taking stock	1
Chapter 2	Some useful electronic components	9
Chapter 3	Switches	15
Chapter 4	Questions	27
Chapter 5	Electrical pressure	36
Chapter 6	A special relay circuit	41
Chapter 7	Microelectronics and chips	58
Chapter 8	The quad NAND integrated circuit	63
Chapter 9	Logic circuits and truth tables – a summary	76
Chapter 10	More questions	81
Chapter 11	The bistable circuit	84
Chapter 12	Drivers	91
Chapter 13	Problem solving	95
Chapter 14	Electronic control systems	101
Chapter 15	Yet more questions	109
Chapter 16	Coding	114
Chapter 17	The pulser, the astable and the clocked bistable	126
Chapter 18	Counting circuits	136
Chapter 19	The latch and the dual decade counter	145
Chapter 20	Memory	155
Chapter 21	Why use digital electronics?	164
Chapter 22	Some final questions	171
Appendix A	Switch contact bounce	176
Appendix B	How to switch a digital clock from a 50 Hz to a 60 Hz input	179
Appendix C	A programmable logic gate	180
	Index	182

Acknowledgements

We are grateful to the Independent Schools' Microelectronics Centre, based in Westminster College, Oxford, for permission to reproduce material which was developed there. This book incorporates much that was included in their series of booklets:

Electronics – a systems approach
Electronics 11–13: Introduction
Electronics 11–13: Experimental Work
Electronics 11–13: Teaching Notes
Electronics 11–13: Construction Notes
Electronics 13–16: Introduction
Electronics 13–16: Worksheets
Electronics 13–16: Teachers' Guide
Electronics 13–16: Projects
Electronics 13–16: Technical Details

We are grateful to the following for permission to reproduce photographs: Barclays Bank, 3.28; Crown copyright, reproduced with the permission of the Controller of Her Majesty's Stationery Office, 7.1; Ferranti Archives, 7.6, 7.7 and 7.8; IBM UK, 7.2 and 7.9; Possum Controls, 14.14; Sanyo, 16.3 (c); Singer, 11.11; TI Creda, 16.3 (b).

All other photographs were taken by Longman Photographic Unit. The electronics modules were lent by Unilab Ltd, Clarendon Road, Blackburn, BB1 9TA, UK.

The cover photograph, by Paul Brierley, is a close-up of a section of printed circuit board showing its gold-plated network.

We are grateful to the following for permission to reproduce copyright material: Cambridge University Press for extracts from *Society and the New Technology*, by Kenneth Ruthren; Victor Gollancz for extracts from *The Mighty Micro*, by Christopher Evans; Her Majesty's Stationery Office for extracts from *The Challenge of the Chip*; Hutchinson Education for an extract from *Microelectronics: A Practical Introduction*, by R. A. Sparkes.

Preface

It has become increasingly accepted that some electronics should be incorporated into physics courses, because electronics has become an important part of modern life and pupils should be introduced to something which will increasingly influence their futures. The National Criteria for Physics for the GCSE examinations state that it is normally expected that physics syllabuses will include some awareness of electronic devices. This raises questions as to what should be taught and how it should be taught.

The Independent Schools' Microelectronics Centre (ISMEC) set about developing a teaching scheme, together with the necessary apparatus, experimental details and guides for teachers, for use in the age ranges 11–13 and 13–16. Extensive trials were held in schools both in the independent and the maintained sectors, as a result of which modifications were made. The final material published by ISMEC is the basis for this book. Apparatus for teaching all of the material has been produced by the major UK manufacturers of science teaching equipment.

An earlier book, **Electronics 11–13**, also published by Longman, covered the ISMEC 11–13 material, which became a compulsory part of the science syllabus of the Common Entrance examination. Since a great deal of the 11–13 material has also been incorporated in certain GCSE syllabuses, it was sensible to produce a new book incorporating much that appeared in the earlier book, together with the extra 13–16 material. This book therefore includes material which can be used in a wide range of courses.

Electronics work of earlier eras might have begun with the characteristic curves of diodes, triodes or transistors. However there is now widespread acceptance that a systems approach is the one which should be adopted, and the emphasis is on this throughout the book. A particular feature is the way it develops from familiar ideas about simple circuits – through switching and the reed relay to the NAND relay circuit – to the quad NAND integrated circuit. There may be some teaching advantage in making the introduction to the NAND gate through the NAND relay, as it avoids introducing something not immediately explicable. Those teachers who prefer to omit reference to the NAND relay will find a satisfactory route through the book omitting Chapter 6. There will be others who will not want to take the work much further than the NAND gate and the bistable, and a suitable route for them could be Chapters 1–7, followed by 13 and 14. Some may not want to go further than Chapter 15, whereas others working with more able pupils will happily move on to the end-point, which is an introduction to memory and its uses. This opens the way to subsequent work on the microprocessor and the computer. There is therefore much flexibility in how the book is used.

Although there are references to analogue electronics, the emphasis is almost entirely on digital work. Digital electronics is the dominant form today, and analogue systems have been increasingly displaced by digital systems as the microelectronics revolution has proceeded. The development of inexpensive microprocessor and memory integrated circuits in the early 1970s generated that revolution – and these of course are digital systems. This rationale, together with the need to resist the temptation to include too much material, is the reason for the emphasis on digital work.

The digital work chosen opens the way to a large number of applications and projects, and these appear throughout the whole book. The criteria for GCSE examinations require that applications should pervade courses and the many projects proposed will certainly show how electronics can be put to use.

The philosophy of the examinations also requires that able pupils too should have the opportunity to show what they know, understand and can do. Some of the material at the end of the book provides scope for their abilities; but it can of course be omitted.

Chapter 1 **Taking stock**

In this book you will be learning about electronic devices and the ways in which they are used, but first, in this chapter, we will remind you of some of the things you need to know about electric currents.

Experiment 1.1   Using a cell and a lamp

Fig.1.1 shows a lamp and a cell connected so that the lamp lights.

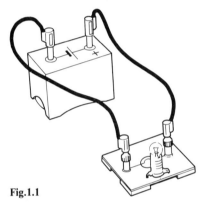

Fig.1.1

1. What happens if you put the cell the other way round in the circuit? Does it make any difference?

2. Does it make any difference if you put the lamp the other way round?

3. Does it make any difference to the brightness of the lamp whether you use long or short leads when connecting the circuit?

4. What happens if there is a gap in the circuit?

The electric circuit

Experiment 1.1 shows that it does not matter which way round you put the cell or the lamp as long as they are joined in a circuit. The lamp will not light if a gap is left. It is usual to speak about an electric current *flowing* round an electric circuit.

Experiment 1.2 Using several lamps and several cells

1. If two lamps are joined in line as in Fig.1.2, they are said to be 'in series'. How does the brightness compare with the brightness produced when the cell was connected to only one lamp?

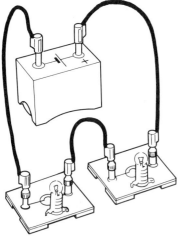

Fig.1.2

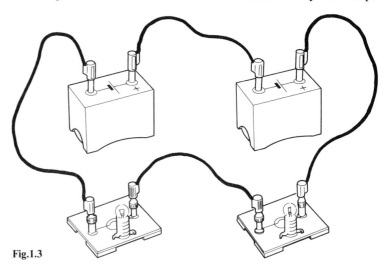

Fig.1.3

2. What happens to the brightness if two cells are connected across the two lamps as in Fig.1.3?

3. What happens to the brightness if one cell is across three lamps in series?

4. What happens to the brightness if there are three cells in series across one lamp?

Brightness of lamps

Experiments show that the lamps glow with the same brightness if there is one cell across one lamp, two cells across two lamps or three cells across three lamps. If more cells are added, the lamps glow more brightly. With fewer cells, they would be less bright. This suggests that three cells drive the same current through three lamps in series, as two cells do through two lamps and as one cell does through one lamp. If the number of lamps is kept the same, adding more cells makes the current bigger.

Measuring current

Current is measured with a meter called an *ammeter*. An ammeter has to be connected in a circuit so that the current to be measured passes through it. The unit in which current is measured is called the *ampere*. When writing down the size of a current, it may be written as, say, 0.3 ampere or 0.3 A.

An ammeter has to be connected in a circuit the right way round. To help you to do this, one of its terminals is coloured red (or marked +) and the other black (or marked −). The ammeter should be connected so that the current, flowing from the positive terminal of the cell (the small central button of the cell), enters the meter through the red terminal.

Experiment 1.3   Measuring current in a series circuit

1. Put two lamps in series with two cells. Connect an ammeter in the circuit as in Fig.1.4, and measure the current.

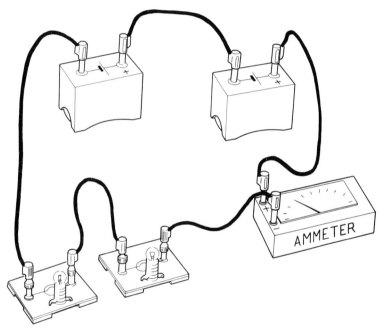

Fig.1.4

2. Now measure the current between the lamps. Also measure it between the cells. What do you notice?

3. Measure the current in a series circuit made from one cell and two lamps.

In a series circuit, the current is the same all round the circuit. Reducing the number of cells results in a smaller current so the lamps are not as bright as normal.

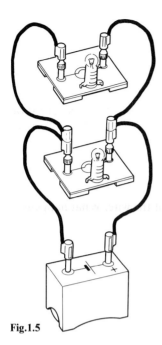

Fig.1.5

Experiment 1.4 Lamps in parallel

When lamps are arranged as in Fig.1.5, they are 'in parallel'. The lamps are placed side by side and joined at each end.

1. Put two lamps in parallel with one cell across the ends. Both lamps should glow with normal brightness.

2. Connect an ammeter in the circuit to measure the current from the cell (Fig.1.6). Make a note of the current.

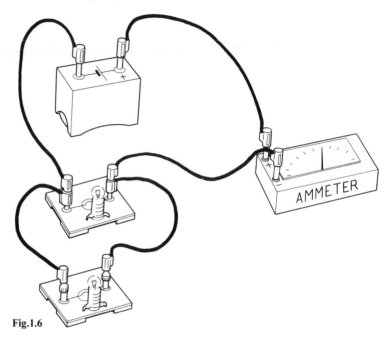

Fig.1.6

3. Now measure the current passing through each lamp and write down the values. What do you notice about these values and the size of the current from the cell?

In a parallel circuit, the current flowing from the cell is equal to the sum of the currents in the parallel branches.

Experiment 1.5 Conductors and insulators

If there is an air gap in a circuit, no current flows. We say that air is an *insulator*. The leads you have been using allow a current to flow: these are therefore called *conductors*.

1. Fig.1.7 shows a cell, a lamp and a piece of paper held between two crocodile clips, all in series. Connect the circuit and find out if paper is a conductor or an insulator.

2. Collect as many different objects as you can and test each of them. Make a list showing which are conductors and which are insulators.

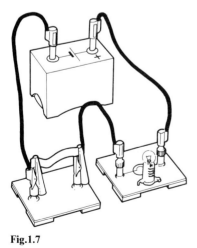

Fig.1.7

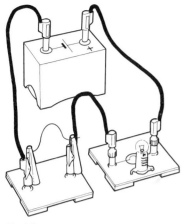

Fig.1.8

Experiment 1.6   Resistance wire

For this experiment you are provided with some special wire: Eureka wire is the name for it, but there is no need to remember that.

1. Use the same apparatus as in Experiment 1.5. Put a very short length of the wire between the crocodile clips (Fig.1.8). Does the lamp light or not?

2. Next try a longer piece of Eureka wire between the crocodile clips. Is there any change in the brightness of the lamp?

3. Now try a very much longer piece of the wire. What happens to the lamp now? If your piece of wire is very long, see it does not touch part of itself. What happens if it does?

The dimmer

The last experiment shows that when a short length of Eureka wire is used, the lamp glows brightly. But when a longer piece is used, it is harder for current to flow and the lamp becomes dimmer. A long wire has a greater resistance than a short wire, and so less current flows. For that reason such wire is called *resistance wire*.

With a short length of resistance wire in the circuit, the lamp is bright. With a long piece the lamp is dim. This suggests that a length of such wire included in a circuit would make a good *dimmer* for changing the brightness of lamps. But a long piece of wire would be inconvenient, so manufacturers wind it up into a convenient coil with a sliding contact and sometimes a control knob on top to turn it (Figs 1.9 and 1.10). This varies the length of the part of the wire through which the current flows.

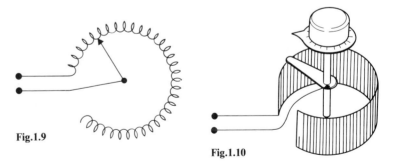

Fig.1.9

Fig.1.10

A piece of wire or other substance which offers some resistance to a current is called a *resistor*. A dimmer is called a *variable resistor* and another name used for it is a *rheostat*.

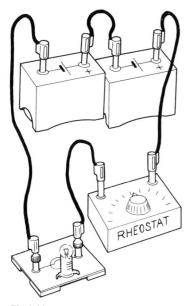

Fig.1.11

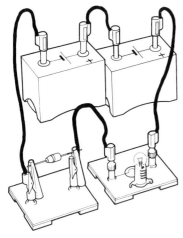

Fig.1.12

Experiment 1.7   The dimmer

1. Connect two cells and a lamp in series so that the lamp glows brightly.

2. Insert a dimmer in the circuit so that you can control the brightness of the lamp (Fig.1.11). Does it make any difference which side of the lamp you put the dimmer?

Experiment 1.8   The diode

1. Connect two cells, a lamp and a diode held between two crocodile clips as shown in Fig.1.12.

2. What happens to the lamp?

3. Turn the cells round. What happens to the lamp this time?

4. Now turn the diode round. What happens?

5. Measure the current in the circuit when the lamp is glowing and when it is not glowing.

The diode is a device which has a low resistance to current flowing one way and a very high resistance to current flowing the other way. In effect current can pass through it in one direction but not the other.

Sometimes it is called a *rectifier*, but the more usual name for it is a *diode*. We shall call it a diode in this book.

Resistance

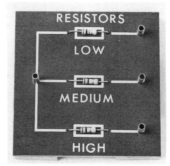

Fig.1.13

Resistance is measured in units called *ohms*. Fig.1.13 shows a module you will use later. It has three resistors labelled high, medium and low. The values of the resistors are about 27 thousand ohms ('high'), 2.7 thousand ohms ('medium') and 270 ohms ('low').

The diode in the previous experiment has a resistance of several thousand ohms when connected one way round, but only about an ohm the other way round.

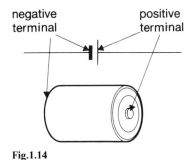

negative terminal   positive terminal

Fig.1.14

Circuit diagrams

To draw a picture of three lamps in parallel across two cells would be a very awkward business if it had to be done in the same way as the circuits shown so far. For this reason scientists use special signs and symbols when drawing diagrams of electric circuits.

The symbol for a cell is two parallel lines as shown in Fig.1.14, one longer and thinner than the other. We have already seen that it matters which way round a cell is used in a circuit. You will notice that the terminals of most cells are labelled + and −. With your cells, the central 'button' is the positive terminal and the metal base at the other end is the negative terminal. In the symbol for the cell, the long line represents the positive terminal, the short line the negative one.

Fig.1.15

Fig.1.16

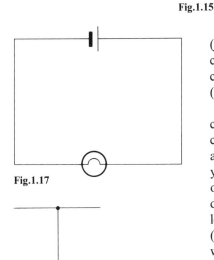

Fig.1.17

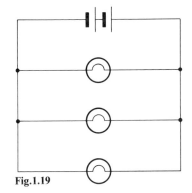

Fig.1.18

When a number of cells (say, three) are connected in series (joined + to − in a line) to form a battery, they can be drawn in a circuit diagram in either of the ways shown in Fig.1.15. Several cells in series are sometimes shown with dashes betwen two cells (Fig.1.16).

The standard symbol for a lamp is shown in Fig.1.17. In this circuit one cell lights one lamp. Straight lines are used to represent connecting leads: they are usually drawn straight, with right-angled corners, to make the diagrams neat and easy to follow. In your experiments, it was necessary to have a complete circuit in order to get lamps to light. A complete circuit is shown in a circuit diagram by joining cells and lamps with lines as in Fig.1.17. Where leads are joined together, the junction is marked with a dot (Fig.1.18). A circuit with three lamps in parallel across two cells would be drawn as in Fig.1.19.

In a circuit diagram the symbol for a resistor was at one time a zig-zag line, but it is now more usual to use the rectangular symbol (Fig.1.20). The symbol for a variable resistor or rheostat uses the same symbol with an arrow through it (Fig.1.21). Diodes are usually shown by the symbol in Fig.1.22. The current flows easily in the direction of the arrowhead, but not the other way round.

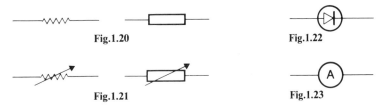

Fig.1.20   Fig.1.22

Fig.1.21   Fig.1.23

The symbol for an ammeter is shown in Fig.1.23.

7

Background reading

A few thousand years ago information was passed from one person to another by word of mouth. The amount of information that a man could have was limited by his memory. Most people had to work very hard just to feed, clothe and house themselves. With the invention of writing and printing people were able to learn new ideas quickly. They learned better ways of growing crops and making tools. They eventually learned how to make machines which worked faster than humans. So fewer people were needed to produce more food and life became a little easier.

Today we have computers to run the machines for us. They also have large memories to store vast amounts of information and more and more people have access to this information. Microelectronics holds the promise of a better life for everyone, but we need to learn quickly how to control microelectronics before it controls us. The more we understand about what microelectronics can and cannot do, the more we shall be able to use this gift properly. This is just one of the arguments for giving microelectronics a place in everyone's education.

(From *Microelectronics: A Practical Introduction*, by R. A. Sparkes.)

1. How do you think people communicated before the days of written material?

2. You have heard about computers. In what ways do you think they make communication easier?

3. What does the author mean about computers controlling us?

4. What do you think might be the reasons for studying microelectronics in schools?

Chapter 2 **Some useful electronic components**

In this chapter we shall consider some devices which play a part in modern electronics and which will be useful to us throughout this book.

The light emitting diode (LED)

The Light Emitting Diode, or 'LED' as it is usually called for short, is an inexpensive device widely used in electronic circuits in order to show that a current is flowing. The circuit diagram symbol for an LED is shown in Fig.2.1.

As an LED can be damaged if too big a current flows through it, the LED module (Fig.2.2) has a resistor in series with the LED and this prevents the current from getting too large. In all the circuit diagrams which follow, that resistor is always shown.

Experiment 2.1 The light emitting diode (LED)

1. Connect an LED module and a battery as in Fig.2.3.

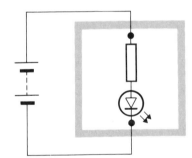

Fig.2.3

2. What happens to the LED when the circuit is connected?

3. Change round the connections to the LED module. What happens this time?

An LED will allow current to pass through it in only one direction. As with an ordinary diode, the arrowhead in the circuit diagram symbol points the way in which current can flow. But the LED differs from the ordinary diode since the passage of current through it causes light to be given out.

Fig.2.1

Fig.2.2

Experiment 2.2   Brightness and current

1. Connect the circuit shown in Fig.2.4, using the LED module, the resistor module and a battery. The resistor module has separate connections so that the resistance can be high, medium or low. Use the low value first. Notice how brightly the LED glows.

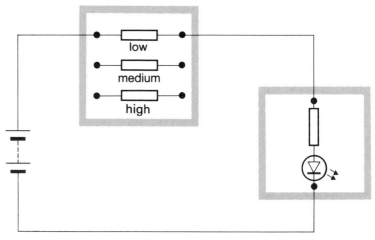

Fig.2.4

2. Now use the medium resistor in place of the low one. What happens to the brightness of the LED?

3. Finally use the high resistance. What happens to the brightness this time?

4. As the resistance gets greater, so the current in the circuit gets less. How does the brightness of the LED depend on the current passing through it?

Experiment 2.3   LEDs in parallel

1. Set up the circuit shown in Fig.2.5, using one red LED module, one green LED module and a battery. Why do both LEDs glow?

2. If another LED were connected in parallel with the above two, how brightly would you expect it to glow? Test your answer using another LED module.

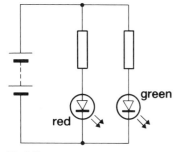

Fig.2.5

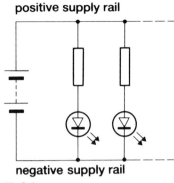

positive supply rail

negative supply rail

Fig.2.6

Positive and negative supply rails

When drawn as in Fig.2.5, the top and bottom lines connected to the battery are known as the positive and negative supply rails. They are labelled as this in Fig.2.6.

If an LED is connected between the positive supply rail and the negative supply rail, a current will flow through it (assuming, of course, that it is connected the right way round).

Parallel connections are often used in electronic circuits. Components connected between the supply rails are in parallel.

Project   A current direction indicator

In this book we will suggest a number of projects from time to time which you may like to try to solve for yourself, though of course you can get advice from your teacher if necessary. However it is the normal job of an electronics engineer to solve problems and it is much more fun to do it on your own.

The problem here is to use a red LED module and a green LED module to construct a current direction indicator. It should be such that when the battery is connected one way round the green LED glows; and when the battery connections are the other way round the red LED glows.

The light dependent resistor (LDR) and the buzzer

The **L**ight **D**ependent **R**esistor (or LDR for short) has the circuit diagram symbol shown in Fig.2.7. The LDR module is shown in Fig.2.8.

Fig.2.7

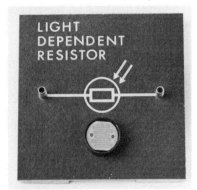

Fig.2.8

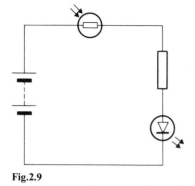

Fig.2.9

Fig.2.10

Experiment 2.4   The light dependent resistor (LDR)

1. Set up the circuit shown in Fig.2.9 using the LDR module in series with an LED module and a battery.

2. What happens to the brightness of the LED when light is shone on the LDR? What happens when the LDR is covered up?

3. What does this tell you about the resistance of the LDR in the light and in the dark?

4. Does the circuit behave any differently if the connections to the LDR are changed round?

The experiment shows that current will flow in either direction through the LDR, as with an ordinary resistor. However, when it is dark, the LDR has a high resistance and therefore allows little current to pass. In bright light, the LDR's resistance falls to a low value and a much bigger current can flow. The resistance of the LDR is perhaps a million ohms in the dark, falling to about 100 ohms in a bright light.

Experiment 2.5   The buzzer

A buzzer makes a sound when an electric current passes through it. It is represented in circuit diagrams by the symbol shown in Fig.2.10. The buzzer module is shown in Fig.2.11.

Fig.2.11

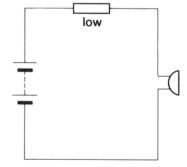

Fig.2.12

1. Set up the circuit shown in Fig.2.12 using the buzzer module, the resistor module and a battery.

2. Investigate whether it makes any difference which way round the buzzer is connected in the circuit.

3. Replace the low value resistor with the medium and then the high value resistors. What effect does this have on the operation of the buzzer?

Project A very simple burglar alarm

The problem is to use an LDR module, a buzzer module and a battery to construct a circuit which will sound an alarm when a light is switched on. If a burglar were foolish enough to turn on the light in a room he had entered, or if the light from his torch fell on the LDR, the circuit could be used to warn the householder.

The motor

The symbol used in this book for a motor in a circuit diagram is shown in Fig.2.13.

Fig.2.13

Experiment 2.6   The motor module

1. Connect the motor module (Fig.2.14) to a battery as in Fig.2.15. What happens when the current flows?

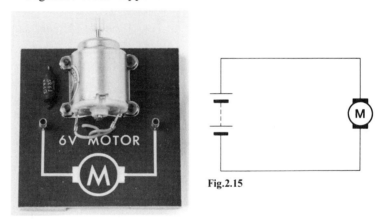

Fig.2.14

Fig.2.15

2. Reverse the battery connections to the motor. What difference does this make?

The above experiment shows that the motor can be driven in either direction depending on the direction in which the current flows through it. It is not always easy to see which way the motor on the module is rotating. You could add a small propeller made from thin card to help. An interesting extension to this experiment would be to add a green LED and a red LED to the circuit in such a way that the red LED lights when the motor rotates one way and the green LED lights when it rotates the other.

Background reading

Microelectronics and the motor car

Microelectronics is about to play a large part in the motor car. It is now possible to buy cars with computer-controlled carburation, which is extremely efficient. It will not be long before it will be in all family cars.

Speedometers which compute your average speed and fuel gauges which 'tell' you in synthetic speech when your petrol is low will be around shortly, and so will headlights which come on automatically when the daylight falls below a certain level.

Another useful electronic device will be a microprocessor which computes the speed of the car travelling in front and assesses whether the two vehicles are being driven at a safe distance apart.

This kind of thing will, presumably, encourage more careful driving, once motorists have got used to the idea of being ticked off by their own cars. But vehicles which pepper one with spoken warnings and instructions may never be popular – a horrific new breed of back-seat drivers.

(Based on *The Mighty Micro*, by Christopher Evans.)

Chapter 3 **Switches**

In this book we shall consider a number of different types of electrical switch. Switching plays a very large and important part in modern electronics. Computers, for example, use electronic switches of various kinds, and the fact that computers can be used to control robots or aircraft is a result of their ability to switch things on or off.

The simplest switch is an on/off switch. The symbol for such a switch in a circuit diagram is shown in Fig.3.1. We know that a current will not flow in a circuit which has a gap in it: when the switch is open, there is a gap, but when it is closed the circuit is completed and a current flows.

The push-button switch (Fig.3.2) is a type of on/off switch such that when the button is pressed, contact is made. The symbol used for it in this book is shown in Fig.3.3.

Fig.3.1

Fig.3.2

Fig.3.3

Experiment 3.1 **Circuits with switches**

1. Connect a battery, an LED module and a push-button switch in series so that you can switch the LED on or off.

2. Now connect the battery, two LED modules in parallel and one switch so that you can switch both LEDs on or off at the same time.

3. Finally, connect the battery, two LED modules and two switches so that one switch operates one of the LEDs and the other switch operates the other LED.

4. When you have arranged your apparatus to work as described in 3, draw a circuit diagram of it.

Questions for homework or class discussion

1. Describe what will happen to each of the lamps in the circuits in Fig.3.4 when the switches are closed (they are drawn in the open position).

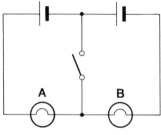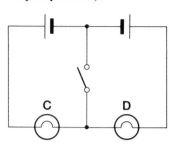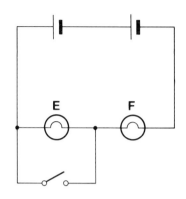

Fig.3.4

2. The circuit in Fig.3.5 shows one cell, two similar lamps (X and Y) and two switches A and B (shown in the open position). When the cell is connected across one lamp it glows with *normal brightness*. Copy the following table and describe whether, on each occasion, the lamp will be brighter than normal, normal, dim or out.

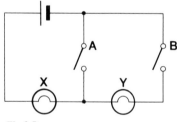

Fig.3.5

Switch positions	Lamp X	Lamp Y
A open, B open		
A open, B closed		
A closed, B open		
A closed, B closed		

3. Describe the effect of opening and closing the switches in each of the circuits in Fig.3.6.

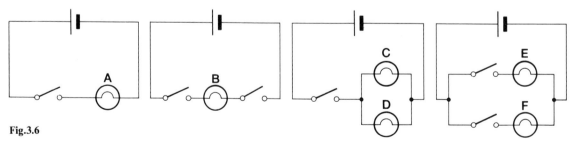

Fig.3.6

16

Experiment 3.2   The push-button switch

This experiment is the same as Experiment 3.1 part 3. If you did
not succeed earlier, try it again. Use a battery, two push-button
switches and red and green LED modules (Fig.3.7).

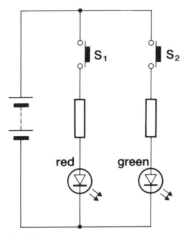

Fig.3.7

1. What happens when S_1 is pressed? And when it is released?

2. What happens when S_2 is pressed?

3. What happens when both are pressed?

The pressure pad as a switch

The pressure pad in Fig.3.8 is another form of on/off switch, in which
the contacts are normally open (the two leads coming from it are not
connected). As soon as any pressure is applied to the pad (by stepping
or sitting on it), the switch is closed and the leads are connected.

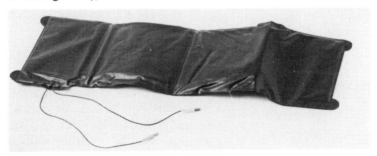

Fig.3.8

Projects with a pressure pad

1. Arrange a pressure pad, buzzer and battery so that the buzzer
   sounds when someone sits down (hide the pad under a cushion).

2. Put a pressure pad under a mat outside the classroom door so
   that a warning is given when your teacher is approaching.

Experiment 3.3 Simple AND circuit

1. Connect two push-button switch modules in series with an LED module and a battery, as in Fig.3.9.

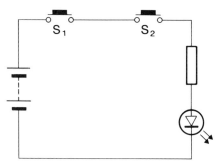

Fig.3.9

2. What happens when S_1 alone is pressed?

3. What happens when S_2 alone is pressed?

4. What happens when both switches are pressed at the same time?

5. Why do you think this is called an AND circuit?

Experiment 3.4 Simple OR circuit

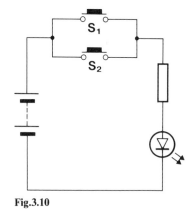

Fig.3.10

1. Using the same modules as in the last experiment, set up the circuit shown in Fig.3.10.

2. What happens when S_1 alone is pressed?

3. What happens when S_2 alone is pressed?

4. Why do you think this is called an OR circuit?

Uses of AND and OR circuits

There are many situations in which an AND circuit might be useful. For example, in a motor car we might want a 'ready to start' light to come on only when the driver has fixed his seat belt AND closed his door. In a bank, it would be a useful precaution if the door to the strong room could be opened only when a switch beside the door AND a switch on the manager's desk were pressed at the same time. Can you think of any more?

An obvious use for an OR circuit is a simple burglar alarm. A pressure pad is placed under a carpet near a door so that the switch is closed by the pressure of the intruder's foot. The OR circuit could be used to protect two doors. An alarm would sound if entry were through either door 1 OR door 2. Can you think of any more uses for an OR circuit?

Fig.3.11

More switches

A simple on/off switch, whose symbol is shown in Fig.3.11, is sometimes called an SPST switch. The letters SPST stand for 'Single Pole Single Throw'. There is one moving contact (the pole) and one position where it makes contact (the throw).

Fig.3.12

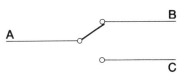

Fig.3.13

Included with your electronics apparatus there are SPDT switch modules (Fig.3.12), where SPDT stands for 'Single Pole Double Throw'. The symbol for an SPDT switch is shown in Fig.3.13. When the switch is in one position, the lead A is connected to B; when in its other position, lead A is connected to C. It is sometimes called a *change-over* switch.

Yet another type (though not included with your apparatus) is the DPDT switch, standing for 'Double Pole Double Throw'. It is like two SPDT switches linked so that they switch over together, as shown in Fig.3.14.

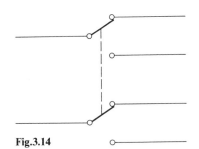

Fig.3.14

Project Manual control of 'stop-go' traffic lights

A *single* set of 'stop-go' traffic lights is to be used to control cars entering a car park through a single lane. The operator sits at one end of the lane to control the lights so that, when a car is leaving, other cars are stopped from entering the lane.

The operator could use the circuit of Fig.3.7 on page 17 to control the lights, but this would not be very satisfactory. Why not?

Instead, use a red and a green LED module together with an SPDT switch and a battery to show how a more satisfactory system could be constructed.

In practice, it would be better if there were *two* sets of lights, each having a green and a red lamp. Devise a circuit which will allow cars to pass through the lane safely from either end. You will need two red LED modules, two green LED modules, an SPDT switch and a battery.

19

Project   Staircase lighting

The problem is to use two SPDT switch modules together with an LED module and a battery to construct a simple staircase lighting system. Suppose one switch is at the top of the stairs and the other at the bottom. It must be possible to turn the light on or off using *either* switch.

Experiment 3.5   Experiment with a motor

The circuit in Fig.3.15 involves two SPDT switch modules, a battery and the motor module. This is a problem experiment in that you should look carefully at the circuit diagram to decide for yourself what it will do. When you have decided, set up the apparatus to find out if you were correct.

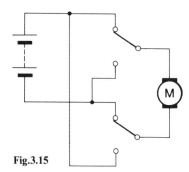

Fig.3.15

The reed switch

Fig.3.16 is a photograph of the reed switch module.

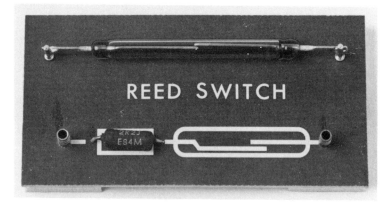

Fig.3.16

Experiment 3.6   Reed switch and magnet

1. Take the reed switch module (Fig.3.16) and examine the two metal contacts inside the glass envelope with a magnifying glass. These contacts are normally open. Put it close to your ear and bring a magnet to the side of the glass envelope. Listen as you do so to see if you can hear a click as the contacts come together.

2. Then connect the switch in series with a buzzer and a battery.

3. Bring a small bar magnet close to the reed switch as shown in Fig.3.17. What happens? What must have happened inside the glass envelope? Use your magnifying glass to see if you are right.

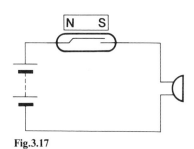

Fig.3.17

20

The reed switch consists of two metal contacts (called 'reeds') inside a glass envelope filled with an inactive (inert) gas to prevent corrosion. Since the reeds are made of a metal containing iron, they can be magnetised by a magnet. If the magnet is brought close to the switch as in Fig.3.18, the metal strips are magnetised, one end being a north pole and the other a south. They are therefore attracted to each other, so that contact is made.

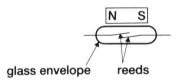

glass envelope   reeds

Fig.3.18

As the magnet is taken away, the strength of the magnetism in the reeds decreases and the springiness of the metal is able to pull the contacts apart again.

In the reed switch module, there is either a resistor or a fuse in series with the switch to make sure that the current is never too big to damage the switch. In the circuit diagrams in this book we will not draw this safety device.

The reed switch described above is an SPST switch: it is either open or closed. We shall later make use of a reed switch which behaves as an SPDT switch. In this switch (Fig.3.19) there is one long reed which acts as the pole of the switch, and two short contact strips. The reed is at first in contact with the lower contact strip, which is made from a non-magnetic material. But when the magnet is brought near, the reed and the top contact strip are magnetised and they come together.

reed   contact strips

Fig.3.19

The reed relay

When you first learnt about electric current, you found that it had both a heating and a magnetic effect. If a current is passed through a small coil, the coil behaves like a magnet.

The reed relay consists of a reed switch like the one you have used, but it is not operated by a magnet. Instead, it has a coil around it as shown in Fig.3.20. If a current passes, the coil behaves like a magnet and, if the current is large enough, the reed switch closes.

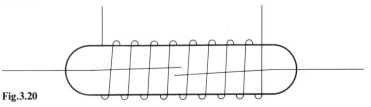

Fig.3.20

In circuit diagrams it would be confusing to draw the coil round the reed switch, so it is usual to draw it at one side, as shown in Figs 3.21 and 3.22.

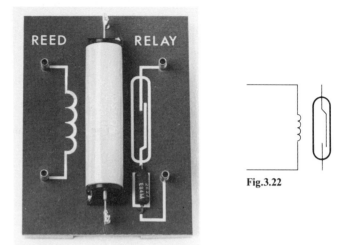

Fig.3.22

Fig.3.21

Experiment 3.7 The reed relay

1. Using the reed relay module, a push-button switch, an LED module and two batteries, set up the circuit shown in Fig.3.23.

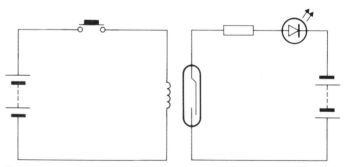

Fig.3.23

2. What happens to the LED when the push-button switch is closed?

This is an important experiment because this time the switching is produced by the flow of an electric current. The two circuits in the experiment are quite separate from each other, but what happens in one is controlled by what happens in the other.

In practice the current through the coil, which is needed to operate the switch, is usually much smaller than the current through the switch contacts. If the LED and resistor were replaced by an electric motor needing a much larger current to make it work, it could be controlled by a much smaller current passing through the coil circuit.

A good example of this is the starter motor of a car. The starter motor may require a very large current (perhaps 100 amperes). This means that the leads from the battery to the motor need to be as short as possible, and, in any case, you do not want large currents of that size going to the dash-board of the car. So a much smaller current is switched on at the dash-board by the ignition switch, and this operates a relay which switches on the much larger current for the starter motor.

Using one circuit to control another plays an important part in electronics. The next experiment illustrates this in motor control.

Experiment 3.8 The reed relay used to control a motor

1. Use the LDR module, the reed relay module, the motor module and two batteries to set up the circuit in Fig.3.24.

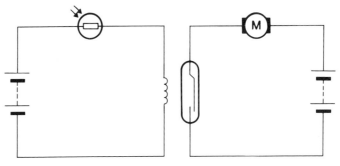

Fig.3.24

2. What happens to the motor when the LDR is covered up? What happens when a light is shone on the LDR?

3. If possible, put an ammeter in each circuit. Is the current in the first circuit very much less than that in the second?

4. You may wonder why it is necessary to use the reed relay at all in the above experiment. Why not use the circuit in Fig.3.25 where the LDR operates the motor directly? Try it and see.

It does not work because the motor needs a large current to operate it and the resistance of the LDR does not decrease enough to make it possible. Even if the resistance of the LDR did fall to a low enough value, it would probably be damaged by the very large current needed to operate the motor.

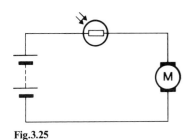

Fig.3.25

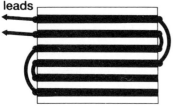

leads

Fig.3.26

Project   Setting up an automatically controlled washing line

A rain sensor can be made with two strips of metal which do not touch each other, as shown in Fig.3.26. No electrical contact is made between the strips until some rain water falls and 'bridges' one of the gaps.

The problem is to use two batteries, the reed relay module, the electric motor module and the rain sensor to make a circuit which switches on an electric motor when the rain falls. Such a circuit could be used for an automatically controlled washing line to bring in the washing when the rain starts to fall!

The experiment will not work with ordinary tap water because your relay is not sufficiently sensitive. However, adding a pinch of salt makes the water an excellent conductor and the necessary complete circuit is provided. (If a real automatically controlled washing line were to be made, a more complicated circuit would be needed to make it sensitive enough to operate the relay when the rain began to fall.)

Experiment 3.9   Controlling a motor with a single power supply

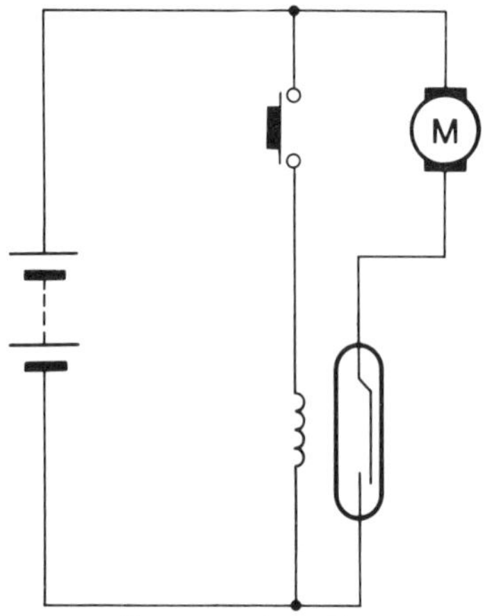

positive supply rail

negative supply rail

Fig.3.27

1. In Experiment 3.8, two batteries were used. In this experiment, only one battery will be used. Set up the circuit (Fig.3.27) with one battery, a reed relay module, a push-button switch and the motor module.

2. What happens to the motor when the switch is closed?

3. In this circuit the relay coil and the relay contacts are connected in parallel. Copy the diagram and mark it with arrows to show the direction of the currents from the battery through each branch of the circuit and back to the battery again.

On page 11, we mentioned that it was useful to think of positive and negative supply rails. The top horizontal line in the circuit is the positive supply rail, and the bottom horizontal line is the negative supply rail. In each of the branches of the circuit the current flows from the positive to the negative supply rail.

Try replacing the push-button switch with an LDR to make a light-controlled motor. As in Experiment 3.8, the current through the coil is much smaller than the current through the motor. The smaller current through the coil branch of the parallel circuit is now controlling the much bigger current through the motor branch.

Background reading

Locks in the future

A familiar old gadget which has been around for hundreds of years is just about to be pensioned off for ever – the key. Already you can obtain electronic locks which open when you punch in the appropriate combination, though they do rely on your remembering what the combination is. Human memory being as fallible as it is, the next development must be a lock which opens only when it has had the chance to scan the electronic chip built into a watch or a signet ring. No one will ever need to hide a key under the mat again.

(From *The Mighty Micro*, by Christopher Evans.)

Fig.3.28

The photograph shows an automatic device by which a bank dispenses money after you present a credit card and type in your code. This is similar to the above.

You could use your electronics modules to design a safety device for a bank vault safe. Suppose a motor is needed to open the door of the vault with one switch outside the door and the other in the manager's office. The motor only works when both switches are pressed at the same time.

Background reading

Controlling the speed of a motor

Lamps, motors and other devices can all be controlled by the output from a computer. Most microcomputers can provide an output voltage of 5 V, and the computer can be programmed so that this voltage is either on or off. In other words, assuming that 5 V is what we call high and 0 V is what we call low, the computer can be programmed so that the output is switched to either high or low. In turn this can be used to control a motor.

The computer can also be used to control the speed of the motor. This may seem surprising at first sight since the program merely switches the 5 V on or off and you might think that, to control the speed of the motor, you want to vary the voltage, and of course that cannot be done when all the computer can do is to switch on or off.

The control is achieved by switching the output high for varying intervals of time. For example, in (a) below, the output is high for more time than it is in case (b).

In case (a), more energy is transferred to the motor than in case (b), and so the motor rotates faster. It is in this way that the speed of the motor is controlled.

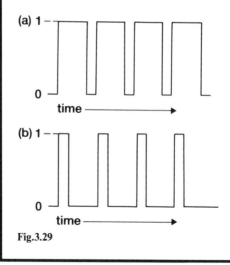

Fig.3.29

Chapter 4 **Questions**

1. Draw a circuit diagram for each of the following:
a   one cell across two lamps and a resistor, all in series;
b   a battery of two cells across two lamps in parallel;
c   a circuit in which there is a current flowing from a battery through a diode, a rheostat and an ammeter in series.

2. Two of the circuits in Fig.4.1 are the same electrically. One is different. Which is the different one?

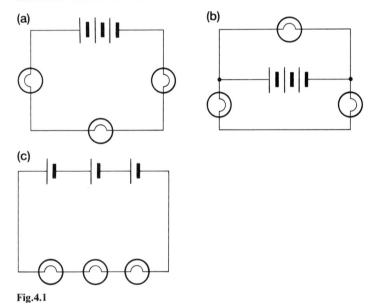

Fig.4.1

3. Are the circuits in Fig.4.2 electrically the same or different? Are the ammeter readings the same or different?

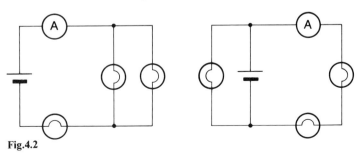

Fig.4.2

4. The circuits in Fig.4.3 include diodes. Which lamps (A, B, C, . . .) will light?

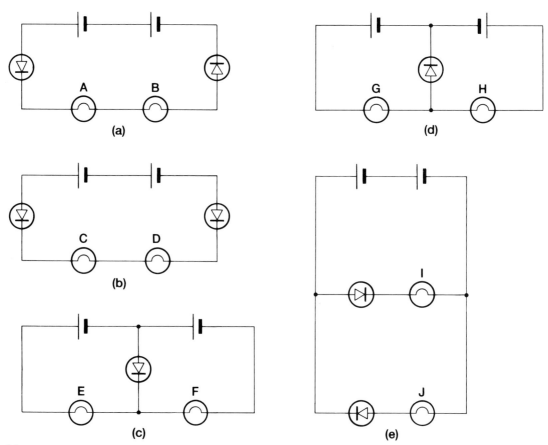

(a)

(b)

(c)

(d)

(e)

Fig.4.3

5. State which lamps (A, B, C, D) in the circuits in Fig.4.4 will have their brightness affected when the rheostats (X, Y, Z) are adjusted.

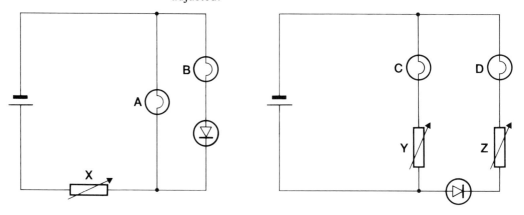

Fig.4.4

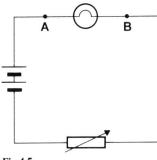

Fig.4.5

6. What would you expect to happen to the lamp in the circuit in Fig.4.5:
a if the resistance of the variable resistor is reduced;
b if a piece of copper wire (a good conductor) is connected between the points A and B;
c if a second, similar lamp is connected to the points A and B in parallel with the first lamp;
d if one of the cells is turned round?

7. For each of the circuits in Fig.4.6, describe how the brightness of the lamps changes as the resistance of the rheostat is reduced.

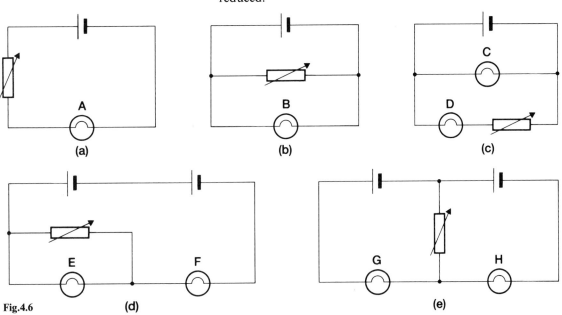

Fig.4.6

(a)   (b)   (c)   (d)   (e)

8. This question compares a current of cars with a current of electricity.
a Two one-way roads P and Q merge into another one-way road R. If one car per minute travels along road P and four cars per minute along road R, how many cars per minute will travel along road Q (Fig.4.7)?
b There is a junction in an electric circuit (Fig.4.8). If 1 A flows along PR and 3 A along QR, what current will flow along RS?

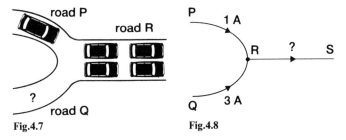

Fig.4.7   Fig.4.8

29

9. All the circuits in Fig.4.9 use similar cells and similar lamps. In figure (a) the current is 0.2 A. In each of the other circuits, say whether the current will be 0, between 0 and 0.2 A, 0.2 A or greater than 0.2 A. Give a reason for each answer.

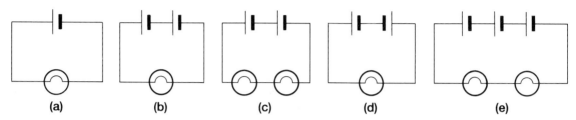

Fig.4.9

10. Describe the effect of closing the switch in each circuit in Fig.4.10. In which circuit will the cell(s) run down most quickly when the switch is closed?

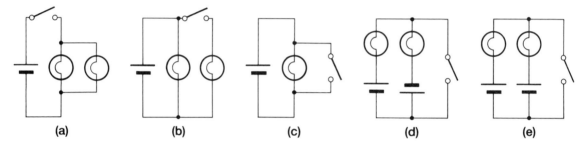

Fig.4.10

11. Each circuit in Fig.4.11 contains a cell, an ammeter, a lamp, a resistor and a switch.
   a   In which circuit does the ammeter measure the current passing through the lamp whether the switch is open or closed?
   b   In which circuit does the ammeter measure the current through the lamp only when the switch is open?

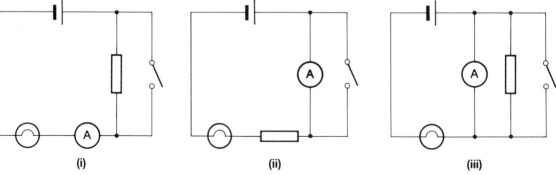

Fig.4.11

30

12. What do the symbols in Fig.4.12 stand for?

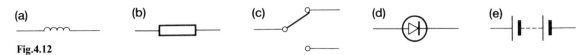

(a) **(b)** **(c)** **(d)** **(e)**

Fig.4.12

13. Draw the circuit symbol used for each of the following:
a an LDR,
b a motor,
c a push-button switch,
d a variable resistor,
e a buzzer.

14. For each of the following, say what the letters stand for and why the component is so called:
a an LED,
b an LDR,
c an SPST switch,
d an SPDT switch.

15. Fig.4.13 is a portrait of Sir Kit Cymbals. How many can you find and what do they represent?

Fig.4.13

16. A reed switch works better if the magnet is held in the position shown in diagram (a) in Fig.4.14 than it does in the position shown in diagram (b). Explain why.

(a) S N   **(b)** N S

Fig.4.14

17. The circuit in Fig.4.15 uses two SPDT switches, A and B.

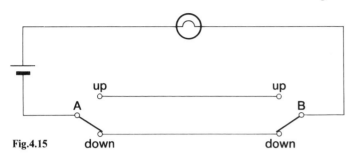

Position of B	Position of A	Lamp
down	down	
down	up	
up	down	
up	up	

Fig.4.15

a The table shows the four possible arrangements for switches A and B, but the column showing whether the lamp is on or off has not been completed. Copy the table and complete it.

b Where are you likely to find a circuit similar to this being used?

18. The table below shows the four possible arrangements for the SPDT switches, A and B, in the circuit in Fig.4.16.

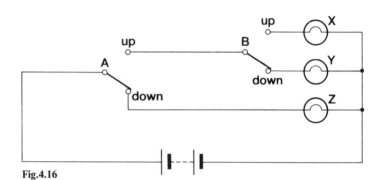

Fig.4.16

Position of B	Position of A	X	Y	Z
down	down			
down	up			
up	down			
up	up			

Copy the table and complete it to show whether each of the lamps, X, Y and Z is on or off.

19. Complete a table (like the one in question 17) for the circuit in Fig.4.17, which contains three switches. Note that there are eight possible arrangements with three switches.

How many arrangements would be possible if a circuit had four SPDT switches in it?

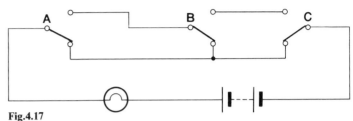

Fig.4.17

20. A room has three doors in it. Near each door, there is a switch which can switch the light in the room on or off, no matter how the other switches are set. The switches needed to do this are shown in Fig.4.18, but the circuit has not been completed. Copy the diagram and complete the circuit.

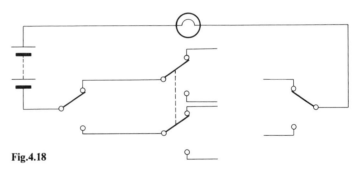

Fig.4.18

21. There are two circuits in Fig.4.19, (a) and (b). What happens, in each circuit, when the switches are operated? Why is (a) a much better circuit to use than (b)?

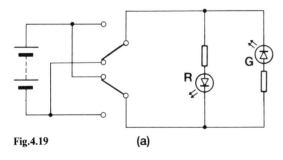

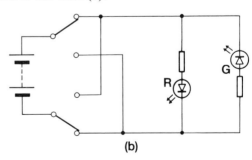

Fig.4.19   **(a)**   **(b)**

22. Two SPST reed switches can be used to make a DPST switch.

a What do SPST and DPST stand for?

b How does a DPST switch differ from two push-button switches?

c How can two reed switches be made into a DPST switch if only one coil is available?

d How can a DPST arrangement be made by using two SPST *relays*, that is, two SPST reed switches each with its own coil? Draw a circuit diagram.

23. In the circuit of Fig.4.20, the light falling on the LDR causes it to have a resistance of 100 ohms, and the ammeter reads 60 mA (60 milliamperes). The light falling on the LDR changes and the ammeter then reads 6 mA.

a Does the resistance of the LDR become greater or smaller?

b Does the light get more or less bright?

c What do you think the new resistance of the LDR is?

24. In the circuit Fig.4.21, the ammeter reads 10 mA.

a What is the colour of the terminal X of the ammeter?

b When switch S is closed, the ammeter reading becomes 30 mA. What is the current at each of the points A, B, C, D, E and F?

c The battery and ammeter are now reversed and the meter reads 28 mA with S closed. Why does the ammeter have to be reversed when the battery is reversed?

d What is now the current at each of the points A, B, C, D, E and F?

e Finally, the non-conducting diode is reversed. What does the ammeter now read?

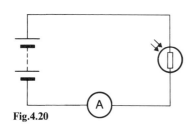

Fig.4.20

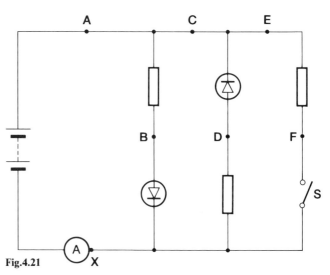

Fig.4.21

25. Draw a circuit diagram of a circuit which will allow you to reverse the direction of rotation of a motor with the aid of two SPDT switches. Include in your circuit a green LED and a red LED so that whichever LED is lit changes when the direction of rotation of the motor changes.

26. In most cars, the interior light comes on when either of the two front doors is opened. Each door operates an SPST switch which is in the open position when the door is closed.
   a Is an OR circuit or an AND circuit needed to do this? Explain your answer.
   b Draw a diagram of a circuit which would work in this way.
   c Add to your diagram another switch which would allow the driver to switch the light on if both doors were closed.

27. A spin drier has a start-stop switch, but the motor will only spin the drum if the lid is closed. Closing the lid closes an SPST switch inside the machine, and the motor can then be started with the start-stop switch.
   a Is an OR circuit or an AND circuit needed? Explain your answer.
   b Draw a diagram of a simple circuit which could be used.
   c Add a buzzer to your circuit diagram so that the buzzer will sound if the lid is not closed and the start-stop switch is operated. (Remember that a buzzer requires much less current than a motor does.)

28. There are 10 errors in the circuit diagram in Fig.4.22. Draw the diagram with the errors corrected.

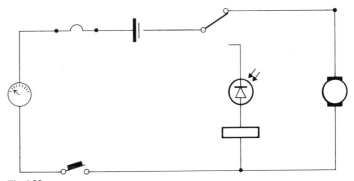

Fig.4.22

Chapter 5 **Electrical pressure**

Suppose you have two syringes, A and B, with a tube fixed between them, and that the space inside is full of water (Fig.5.1).

Fig.5.1 A B

You can make water pass from A to B through the tube by pressing the piston of syringe A. Doing this makes the pressure in A greater than the pressure in B and this *difference* in pressure causes water to flow through the tube. The harder you push on the piston of A the greater is the pressure difference and the faster the water flows, so there is a bigger current. Of course, the current will stop when you stop pushing or when the piston cannot move any further.

If a current is to flow all the time, then a water pump has to be used to keep a difference in pressure between A and B (Fig.5.2).

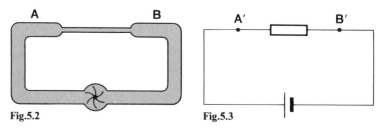

Fig.5.2 **Fig.5.3**

This is just like an electrical circuit in which there is a cell (an electricity 'pump') causing electric current to flow through a resistor (Fig.5.3). The cell pumps electric charge through the circuit, and we call this movement of charge an electric current. The cell causes a difference of electrical 'pressure' between A′ and B′, and the effect of this is an electric current flowing from A′ to B′ through the resistance wire.

If the pipe between A and B in the water circuit is shorter, then the same difference of pressure between A and B causes a larger current, because it is easier to push water through a short pipe than through a long one. And the electrical circuit is like that too – a shorter piece of resistance wire allows a larger current to flow. What do you think would be the effect of using a fatter wire?

There are other similarities between electrical and water circuits.

1. If the circuits have parallel branches (Fig.5.4), some current flows through each branch. For both circuits, the sum of the currents flowing in each branch equals the current flowing from the pump.

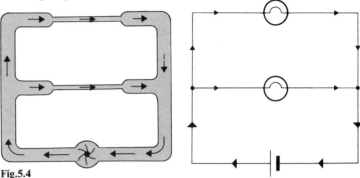

Fig.5.4

2. If one of the pipes had a stop-tap in it, we could use the tap to stop the water current even with the pump still working. The water pressure on one side of the tap would then be bigger than on the other side, but water could not pass through the tap.

   Electrical circuits can be like that if we put a switch in the circuit: there is no current when there is a gap in the circuit, but there is a 'pressure' difference across the switch.

3. Notice, too, that the water pump does not create water! The water is there all the time, and all the pump does is to move it through the pipes. So it is with the electrical circuit. The battery does not manufacture electric charge. The electric charge is in the wires all the time, and all the cell does is to move the charge.

Measuring electrical 'pressure' difference

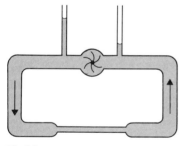

Fig.5.5

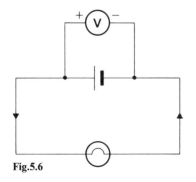

Fig.5.6

In a water circuit we could measure the difference in pressure between the two sides of the pump by inserting a long tube on each side of it (Fig.5.5). On one side of the pump, the pressure would be 'high' and on the other side 'low', and the difference in pressure would be measured by the difference in the levels of the water in the two tubes.

Electrical 'pressure' difference is measured with a voltmeter in units called volts. The circuit diagram symbol for a voltmeter is seen in Fig.5.6. As with the water circuit, the voltmeter has to be connected between the points whose 'pressure' difference we require; that is to say, it is connected in parallel with the cell or *across* it. Notice that this is different from an ammeter which is connected in a circuit in series.

Fig.5.7

Like an ammeter, the voltmeter must be connected the right way round. One terminal is usually coloured red (or marked +) and this should be connected to the positive side of the cell, or the point where the 'pressure' is higher. The other terminal is usually black (or marked −), and this should be connected to the negative side of the cell, or the point where the 'pressure' is lower.

Experiment 5.1   Using a voltmeter

1. Connect three cells in series (Fig.5.7). Measure the 'pressure' difference or voltage between P and Q with a voltmeter. Note its reading.

2. Now measure the voltage between Q and R, and then between R and S. The readings should all be close to 1.5 volts.

3. Then measure the voltage of two cells (between P and R, or between Q and S). Finally, measure the voltage of three cells in series (between P and S).

Each of the cells you have used produces a 'pressure' difference of about 1.5 volts, or 1.5 V. A battery of two cells in series produces twice as much, about 3.0 V, and a battery of three cells gives 4.5 V. Batteries using this type of cell and giving 3, 4.5, 6 or 9 V can be bought. Different kinds of cell can be obtained; one type produces 1.2 V, another 2.0 V, for example. See if you can find out what voltage a car battery gives, and how many cells there are in it.

Putting cells in series to form a battery produces a more powerful 'pump', and more powerful pumps cause more current to pass through a circuit. The voltage from a cell or battery remains fairly steady over most of its life and then falls towards the end of it. A graph of its voltage against time might look like Fig.5.8, though how rapidly it falls depends on the current which flows through it.

voltage

0   time

Fig.5.8

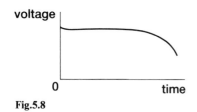

Fig.5.9

Fig.5.10

Cells in parallel

Until now, we have considered only cells in series. The next experiment deals with cells in parallel.

Experiment 5.2   Cells in parallel

1. Connect a voltmeter across one cell and measure the electrical 'pressure' difference it causes (Fig.5.9).

2. Add another cell in parallel with the first, taking care to make sure the cells have their positive terminals connected together (Fig.5.10). What is the voltage now?

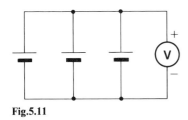

Fig.5.11

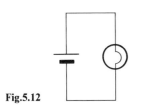

Fig.5.12

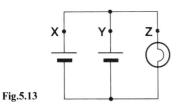

Fig.5.13

positive supply rail

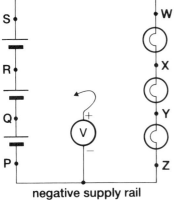

negative supply rail

Fig.5.14

3. Try adding a third cell (Fig.5.11). Does this change the voltage?

Connecting cells in parallel does not change the voltage produced. A lamp connected across the cell arrangements shown in Figs 5.9, 5.10, and 5.11 would light with normal brightness in each case. So, what would be the difference?

4. Connect the circuit shown in Fig.5.12, and use an ammeter to measure the current flowing.

5. Now add a second cell in parallel (Fig.5.13) and measure the current at each of the points X, Y and Z. What do you notice?

You should find that the current through the lamp in each circuit is about the same, but the current through each cell in 5 is only half the value it was in 4. Can you guess the current which would flow through each cell if a third cell in parallel were added?

The advantage of having cells in parallel is that they last longer when each has to push a smaller current.

Voltage levels

Experiment 5.3   Measuring voltage levels

1. Set up the circuit in Fig.5.14, with the black ($-$) terminal of the voltmeter connected to the negative supply rail.

2. Connect the red ($+$) terminal of the voltmeter to P, and note the reading.

3. Connect the red ($+$) terminal, in turn, to Q, R, S, W, X, Y and Z, and note the readings.

In this experiment, you are measuring the electrical 'pressure' difference between a certain point in the circuit and the negative supply rail where the 'pressure' is lowest. You are measuring how high the level at that point is above the lowest level. The level at Q is about 1.5 V, at R about 3 V, at S about 4.5 V. The level drops in similar steps from W to X to Y to Z. At the points P and Z the level is the same as that of the black terminal of the voltmeter, and so the voltmeter gives a zero reading.

In the work ahead, we will often use the idea of voltage levels (with respect to the negative power rail). We shall refer to the level of the positive supply rail as the high level and that of the negative supply rail as the low level.

Of course, if a conductor of any sort is connected between different levels, the electrical 'pressure' difference will cause a current to flow through the conductor. But, if the ends of a conductor are connected to the same voltage level, then no current will flow through it because there is no electrical 'pressure' difference to cause it.

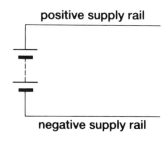

positive supply rail

negative supply rail

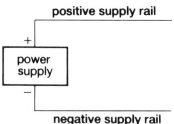

positive supply rail

power supply

negative supply rail

Fig.5.15

Power supplies

In Fig.5.14 it was the battery which gave the different voltage levels. It is possible to produce positive and negative supply rails using any power supply unit, including a mains driven one, provided that the supply output voltage is of the right level for the modules (Fig.5.15).

Background reading

A well-informed society?

We spend less time at work than we did thirty years ago. In thirty years time we shall spend even less. At present, leisure is for many a respite from something they have to do in order to exist; it is a time for rest and for casual amusements and passing sensations. Pessimists argue that when work takes up less time and becomes less demanding we shall become aimless, incapable of expressing ourselves, and dulled by trivial entertainment coming to us mainly through our television sets! Optimists say that the division between work and leisure will disappear. We shall have more chance to exercise our creative and cultural talents, more time to assist in the improvement of society and more incentive to learn about ourselves and the world about us.

It is in providing the means to learn that microelectronics may best serve us. Prestel and personal computers are but the forerunners of systems that will enable us to learn when we want to rather than during prescribed periods, and will provide us with many more sources of information. Microelectronics will create what is coming to be known as the 'information society', a society in which we shall be able to communicate knowledge and experience far more easily than we can now. Five hundred years ago Renaissance culture was disseminated by the newly born printing press. Microelectronics stands ready to spread the culture of today and tomorrow.

(From *The Challenge of the Chip*, HMSO.)

Chapter 6 A special relay circuit

Radios and television sets, record players and tape recorders, digital watches and computers, contain complicated circuits designed by electronics engineers to be useful to us.

In designing these circuits, the engineers frequently use a number of smaller, simpler circuits which are connected together in a suitable way to do what is required. They know how to do this because they are familiar with the ways in which the simpler circuits behave. In other words, they use these simpler circuits as 'building bricks' when constructing more complicated circuits.

In this chapter you will meet one of the most important electronic building bricks. It is usually called a NAND circuit, but that name need not worry you at this stage. First we will find out what it does, then we will see how it works and how it is put to use. Chapter 8 will introduce NAND circuits made in a different way.

The NAND relay module

In order that the NAND relay module (Fig.6.1) shall work, it is necessary to connect it to a battery as shown (Fig.6.2). The positive terminal of the battery provides the positive supply rail, the negative terminal the negative supply rail. As explained in the previous chapter, there is a difference in voltage level between these two rails, and we shall refer to the level of the positive supply rail as the *high* level and that of the negative supply rail as the *low* level.

Fig.6.1

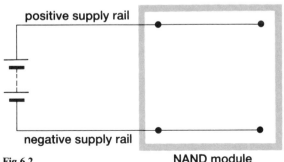

Fig.6.2   NAND module

The NAND relay module has two input sockets, labelled A and B in Fig.6.3, and an output socket. The voltage level at the output socket depends on the voltage levels at the A and B input sockets. To show the voltage level at the output socket, a voltmeter can be connected between it and the negative supply rail as in Fig.6.3.

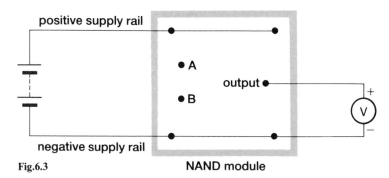

Fig.6.3   **NAND module**

Another way to show the voltage level at the output is to connect an LED module between the output socket and the negative supply rail, as in Fig.6.4. Then, if the voltage at the output is high (that is, at the level of the positive supply rail), a current will flow through the LED to the negative supply rail. If the voltage at the output socket is low (that is, at the level of the negative supply rail), no current will flow and the LED will not light. (Of course, this assumes that the LED has been connected the right way round; if it were the wrong way round, it would not light at all!)

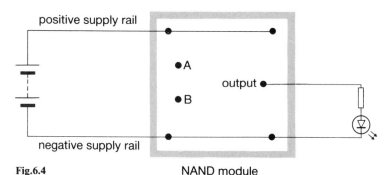

Fig.6.4   **NAND module**

Experiment 6.1   Using the NAND relay module as an inverter

1. Set up the circuit (Fig.6.5) using a NAND relay module, an LED module and a battery.

2. Plug a lead into input socket A. Such a lead, the other end of which is not connected at first to anything, is often called a 'flying lead'. Connect the other end of the lead to the positive supply rail. What does the LED tell you?

3. Now connect the flying lead (still connected to socket A) to the negative supply rail. What does the LED tell you this time?

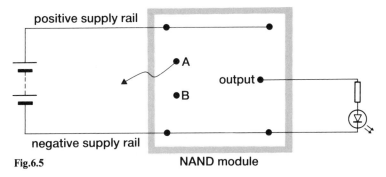

Fig.6.5   NAND module

4. Remove that lead and try the same experiment with the flying lead connected to input socket B. Is the behaviour the same as for input socket A?

5. When the flying lead is not connected to either of the supply rails, does the input behave as if it is high or as if it is low?

You should have noticed that the output is at a high level (LED lit) only when the input level is low, and that both inputs behave in the same way. An unconnected input behaves as though it is at the high level. Such an input is usually said to be 'floating'.

This behaviour can be shown in a *truth table* (Fig.6.6).

Notice that the circuit changes the voltage level over; if you connect the input to the low level (the negative supply rail), the output goes to the high level, and vice versa. This is why the circuit is called an *inverter* ('invert' means 'turn upside down').

The circuit can be operated with a switch or an LDR at the input, and the output can be used to operate a buzzer or motor, as in the next experiment.

Input	Output
low	high
high	low

Fig.6.6

Experiment 6.2   Operating an inverter with an LDR

1. Use a NAND relay module, a buzzer module, an LDR module and a battery to set up the circuit shown in Fig.6.7.

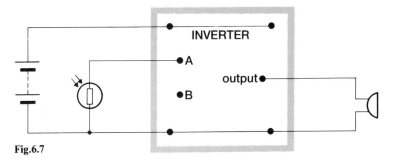

Fig.6.7

2. What happens when you shine the light from a torch on to the LDR?

When light is shone on the LDR its resistance becomes small and it behaves like a *closed* switch. This means that the input is effectively connected to the negative supply rail. It is therefore low and that means the output goes high so that the buzzer sounds.

In the dark the resistance of the LDR is very high and it behaves like an *open* switch. The input is then unconnected (it is 'floating'), and you found in the last experiment that a floating input behaves as though it were high. A high input results in a low output, so the buzzer will be off.

Project   Make an automatic light

Use two NAND relay modules, an LDR module, an LED module and a battery to make a circuit which will light the LED automatically when it gets dark.

Experiment 6.3   Using both the inputs of the NAND module

1. Use a NAND relay module, an LED module and a battery to set up the circuit in Fig.6.8.

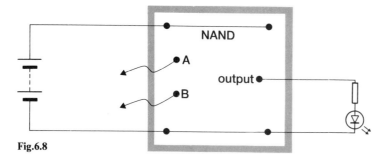

Input B	Input A	Output
low	low	
low	high	
high	low	
high	high	

Fig.6.8

2. Plug a flying lead into each of the inputs of the NAND module.

3. Make a copy of the truth table shown above. Connect each of the input leads to either the positive or negative supply rails in order to complete the truth table, showing whether the output is high or low. Remember that each of the input leads must be connected to either the positive supply rail or the negative supply rail.

Truth table for the NAND circuit

The truth table for the NAND circuit is as shown below.

Input B	Input A	Output
low	low	high
low	high	high
high	low	high
high	high	low

Notice that the output is high whenever one or both inputs is at a low level. You could say that, with a NAND circuit, the output is:

Not high only when input A **AND** input B are high.

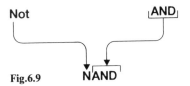

Fig.6.9

That is why it is called a NAND circuit (Fig.6.9).

An AND circuit

An AND circuit would have a different truth table. It would have a high output only if both inputs A **AND** B were high.

Input B	Input A	Output
low	low	low
low	high	low
high	low	low
high	high	high

Experiment 6.4   Making an AND circuit

1. Use two NAND relay modules (one as an inverter), an LED module and a battery to set up the circuit in Fig.6.10.

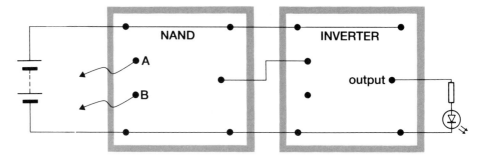

Fig.6.10

Input B	Input A	Output
low	low	
low	high	
high	low	
high	high	

2. Attach flying leads to each of the inputs A and B of the first NAND module. Copy the truth table. By connecting the flying leads to the positive or negative supply rails, complete the truth table. Does your circuit behave like an AND circuit?

3. Compare the truth table for this circuit with the table obtained for the NAND circuit. What has the inverter done?

A convenient shorthand

It becomes a little tedious writing 'high' and 'low' in truth tables. Electronics engineers usually write 1 for 'high' and 0 for 'low'. Thus the truth table for a NAND circuit changes like this:

Input B	Input A	Output		Input B	Input A	Output
low	low	high	becomes	0	0	1
low	high	high		0	1	1
high	low	high		1	0	1
high	high	low		1	1	0

Practise the new shorthand by using it to rewrite the truth tables for the inverter and the AND circuit.

Experiment 6.5 Using a NAND circuit to make a simple burglar alarm

1. Use a NAND relay module, two push-button switches, a buzzer and a battery to connect the circuit shown in Fig.6.11.

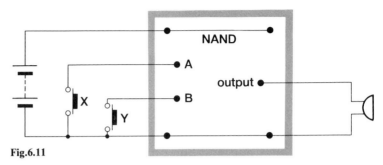

Fig.6.11

2. What happens when switch X or switch Y is pressed? What happens when both are pressed together? Explain why this happens.

3. It makes a more realistic burglar alarm if the push-button switches are replaced by pressure pads.

There is still a serious weakness in this simple circuit for use as a burglar alarm. The alarm stops when the intruder steps off the pressure pad. A better alarm would continue to sound once it had been set off whether or not the switch was released. To make such an alarm, a circuit called a *bistable* is needed. This can be made using two NAND relay modules and will be discussed on page 52.

How the NAND relay circuit works

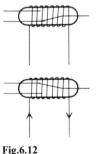

Fig.6.12

On page 19 we described the SPDT reed switch. The NAND relay module uses such a reed switch with a coil round it, as in Fig.6.12.

The reed (the pole of the switch) is normally touching the lower of the two contacts, as shown. When a current flows through the coil, the coil magnetises the reed and the upper contact, so that the reed moves up to touch the upper contact. (The lower contact is made of a non-magnetic material so that it is not magnetised.) The reed stays in the upper position until the current in the coil stops flowing, when it springs back to touch the lower contact again.

How a reed relay works as an inverter

The battery provides the positive and negative supply rails (Fig.6.13).

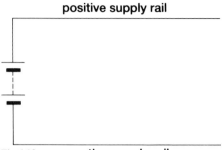

positive supply rail

Fig.6.13   negative supply rail

An SPDT reed switch is connected between the positive rail and the negative rail (Fig.6.14). The pole of the switch is connected to the output point P, and this is normally touching the lower switch contact so that the output is low.

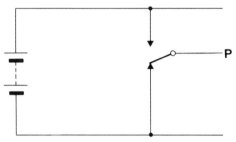

Fig.6.14

Round the reed switch is a coil, making it into a reed relay. In Fig.6.15 it would be confusing to draw the coil round the reed switch, so it has been drawn for convenience at the side. The coil is connected between the positive supply rail and the input A.

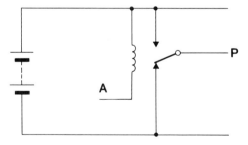

Fig.6.15

What happens if A is connected to the positive rail (Fig.6.16)? As J and K are both at the same voltage level, no current will flow through the coil. The reed switch will not change. P will still be connected to the negative rail. Thus if A is high, P is low.

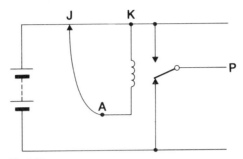

Fig.6.16

What happens if A is connected to the negative rail (Fig.6.17)? The coil is now connected between the positive and negative supply rails, so a current will flow through it from K to L. The current in the coil produces a magnetic field – and the reed switch changes over. P is now connected to the positive rail. Thus if A is low, P is high.

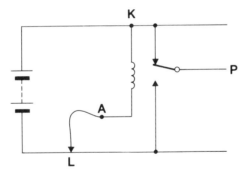

Fig.6.17

The truth table is shown below. We have an inverter.

Input A	Output P		Input A	Output P
low	high	or	0	1
high	low		1	0

How a reed relay works as a NAND circuit

As before, the battery provides positive and negative supply rails and an SPDT reed switch is connected between them, with the pole connected to the output point P. Normally the pole is touching the lower switch contact so the output is low (Fig.6.18).

A coil is again round the reed switch, making it into a reed relay. But this time the coil is connected between the positive rail and *two* input points A and B, as in Fig.6.19.

positive supply rail

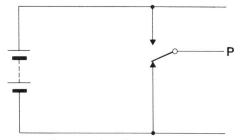

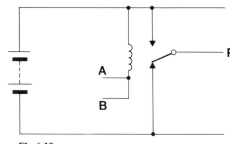

Fig.6.18   negative supply rail

Fig.6.19

What happens if A and B are both connected to the positive rail (in other words, if A and B are both high) as in Fig.6.20? No current flows through the coil and the output therefore remains low. But if A and B are both connected low as in Fig.6.21, a current flows through the coil, the reed switch changes and the output goes high.

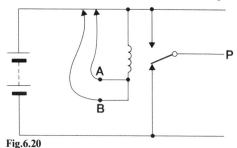

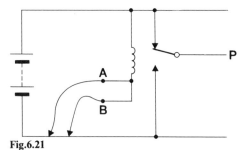

Fig.6.20

Fig.6.21

We have begun to complete a truth table, as follows:

Input B	Input A	Output
0	0	1
0	1	
1	0	
1	1	0

50

What happens if A is high and B is low? A current would immediately flow, but not through the coil. Such connections would merely short-circuit the supply and rapidly discharge the battery (Fig.6.22).

To prevent this happening, a diode is added near the input socket A. Then as B is low, a current will now flow through the coil and the output goes high (Fig.6.23).

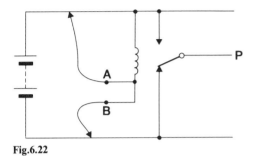

Fig.6.22

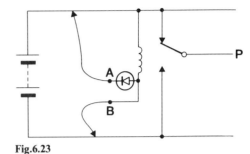

Fig.6.23

What happens when B is high and A is low? Again, the battery is short-circuited, and a second diode near B is needed (Fig.6.24). Now when B is high and A is low, a current flows through the coil and the output is switched to high.

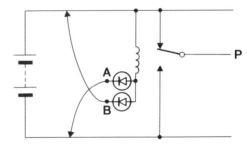

Fig.6.24

We can now complete the truth table:

Input B	Input A	Output
0	0	1
0	1	1
1	0	1
1	1	0

Fig.6.25

and we have a NAND circuit.

The circuit is marked on the front of the NAND relay module (Fig.6.25). Remember, though, that you can use the NAND module to do useful jobs without understanding how it works. In electronics today, even circuit design engineers use building bricks such as the NAND circuit without worrying about how they work. The important thing is to know what the building bricks will do, and how to make use of them.

The bistable

Two NAND modules can be connected to make what is called a bistable circuit. Bistable circuits are used a great deal in modern electronics, especially in computers.

Experiment 6.6   Making a bistable circuit

1. Set up the circuit in Fig.6.26 using two NAND modules, two LED modules and a battery. The two NAND modules have the same positive and negative supply rails.

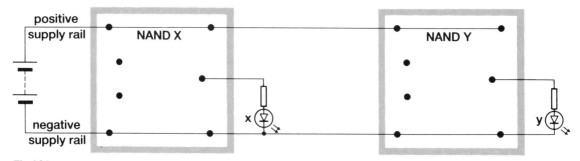

Fig.6.26

The LED modules are connected between the outputs of the NAND modules and the negative supply rail. They will show whether the outputs are high or low.

2. Now add a lead which connects the output of NAND X to one of the inputs of NAND Y (Fig.6.27). What do the LEDs show? (As there are no connections to the inputs of NAND X, the output of NAND X is low, and therefore the input of NAND Y must also be low. And that means that the output of NAND Y must therefore be high and LED y is lit.)

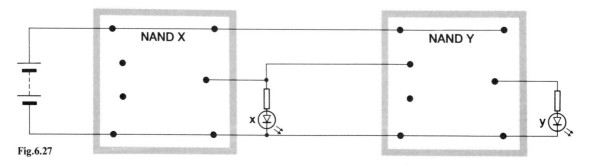

Fig.6.27

3. Next, connect a lead from the output of NAND Y to one of the inputs of NAND X. (Note that in drawing this in Fig.6.28 the lead from NAND Y crosses over the positive supply rail. This does **not** mean that they make electrical contact. If they were connected there would be a dot at the junction.)

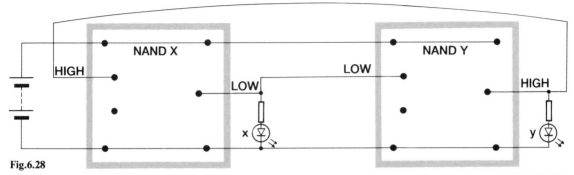

Fig.6.28

When NAND Y's output is connected to the input of NAND X, it makes that input high. This does not change the output of NAND X, which remains low. LED x remains off.

However long the connections are left, LED x remains off and LED y remains on. It is a *stable* arrangement.

4. But this is not the only stable arrangement. Remove the two leads connecting the inputs and outputs of NAND X and NAND Y. Then connect first the output of NAND Y to the input of NAND X and afterwards connect the output of NAND X to the input of NAND Y. You will now have set up the other stable arrangement. How does this second stable state of affairs differ from the first one?

In the first state,   LED x is off,   LED y is on.
In the second state,   LED x is on,   LED y is off.

Because the circuit has two stable states, it is called a *bistable* circuit.

5. Now add two push-button switches to the arrangement of modules, as in Fig.6.29. What happens when you press switch Q? Try pressing it several times.

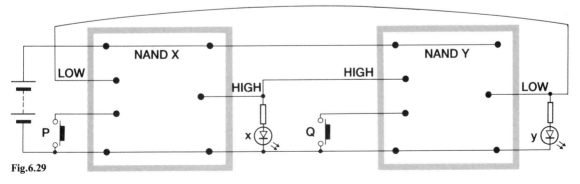

Fig.6.29

53

6. What happens when you press P? Try pressing it several times.

7. Now press Q again, and then P. Then press P and Q one after the other.

Pressing a switch causes the bistable to switch from one of its stable states to the other, *provided* it is the switch connected to the NAND module whose output is low (i.e., the one with the LED not lit).

In the first circuit below (Fig.6.30), it is only when you press Q that the circuit flips over to that shown in the second (Fig.6.31). After it has flipped over, pressing Q again has no further effect.

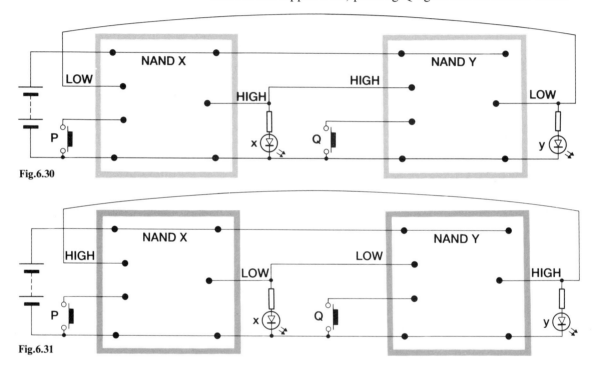

Fig.6.30

Fig.6.31

When Q is pressed for the first time, the input of NAND Y is connected to the negative supply rail and immediately the output of NAND Y becomes high. That gives a high input to NAND X, so that its output becomes low. Because that is connected to the input of NAND Y, NAND Y now has a low input (and therefore a high output) whether Q is pressed or not. The only way to get the bistable to flop back again is to operate switch P. And now you can understand why bistable circuits are often called 'flip-flops'.

Experiment 6.7   A latched burglar alarm

Having built the bistable, you are now in a position to build a really good burglar alarm.

1. First replace LED y with a buzzer. What happens when you switch from one stable state to the other?

2. Now replace the push-button switch at the input of the NAND Y module by a pressure pad. What happens when the pressure pad is pressed?

Applications using a bistable circuit

Project 6(a) A 'teacher detector'

A small variation of the arrangement in Experiment 6.7 will turn it into a 'teacher detector'. You could put a pressure pad under the mat in the corridor outside your classroom to let you know when your teacher is coming.

It would be better if the buzzer could be turned off before the teacher arrives. Perhaps this could be done by putting a second pressure pad under another mat immediately outside the door to change back the bistable when the teacher steps on that mat.

Your teacher will then automatically set off the alarm and silence it before entering the room. We leave that as something for you to try.

Project 6(b) Controlling an electric motor

Use two NAND relay modules (as a bistable circuit) and two push-button switches to control an electric motor so that, in one of the stable states, the motor goes one way round, and in the other state, it goes the opposite way.

Circuit diagrams

The diagrams in this chapter have used a special symbol for the NAND circuit. This was to help you to set up the circuits correctly. You will have noticed that drawing circuit diagrams became more difficult as the circuits became more complicated. The diagrams were harder to understand and they took more space as well. Some simpler way of showing circuits is necessary.

The way this is done is first to replace the NAND relay circuit by a new symbol. This is shown in Fig.6.32.

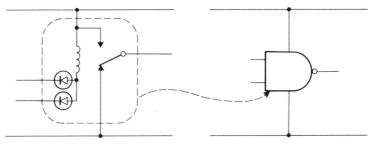

Fig.6.32

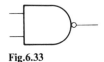

Fig.6.33

The next step is to omit the power supply connections, leaving just the symbol of Fig.6.33.

It is, of course, important to remember that a power supply is needed to make the NAND circuit work. All we are doing is making our circuit diagrams simpler by not actually drawing the power supply connections.

The symbol of Fig.6.33 is the usual one used in electronics for a NAND circuit. The two lines on the left represent the two inputs, and the one on the right the output.

Background reading

Fig.6.34

The transistor switch

The photograph (Fig.6.34) shows a transistor, which is a small electronic component with three leads. These leads are labelled *b*, *c*, and *e* in Fig.6.35, standing for base, collector and emitter, although you do not need to know the reason for these names in order to understand what a transistor does.

A transistor can be thought of as a special kind of electronic switch. It can do a job very similar to that of the reed relay, but it has no moving parts. The circuits in Fig.6.36 show how a transistor and a reed relay can be used to turn on an LED. In either case, if the switch is pressed, the LED lights.

Fig.6.35

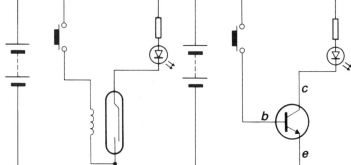

Fig.6.36

The transistor behaves as though there is a switch between its *c* and *e* leads, which is operated by connecting the *b* lead to the positive supply rail (Fig.6.37). When the push-button switch is pressed, the 'SPST switch' between *c* and *e* closes and the LED lights.

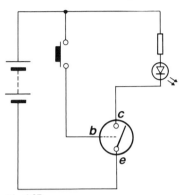

Fig.6.37

This behaviour is very like that of the reed relay. In the case of the relay, the current in the coil circuit controls a current through the reed switch. With the transistor, the *b* lead controls the current flowing between the *c* and *e* leads.

Until recently transistors were very common in electronic circuits. They are small, cheap and very reliable. But nowadays, they have been largely replaced by *integrated circuits*.

Chapter 7 **Microelectronics and chips**

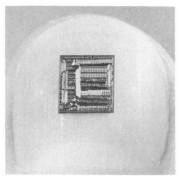

Fig.7.1

Fig.7.2

From television, books and magazines, you may have heard about microelectronics and the so-called 'chip'. What is microelectronics? What are chips? Why are they important?

Microelectronics is about very small electronic circuits. In the experiments you have been doing you have used devices which are quite large. In 1959 scientists found ways of making switching circuits in small, thin layers of silicon. This piece of silicon is known as a *chip*. It can be about the size of the nail of your little finger or even smaller (Fig.7.1).

The important thing was that several switching circuits could be made in the same chip. At first there were 10 or 20, but ways have now been found to make the switching circuits smaller and smaller, and today more than 100 000 can be put in a single piece of silicon only 5 mm square! These circuits can be connected together electrically inside the chip to make very complicated circuits. A magnified view of a silicon chip containing many circuits is shown in the photograph (Fig.7.2).

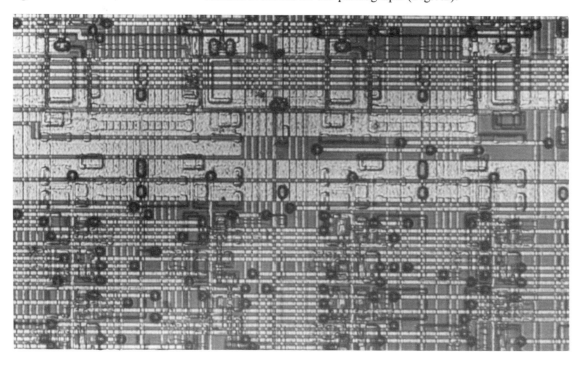

Another name for a silicon chip containing many electronic circuits is an *integrated circuit* (or 'IC' for short). If you buy an integrated circuit, it will probably look something like Fig.7.3.

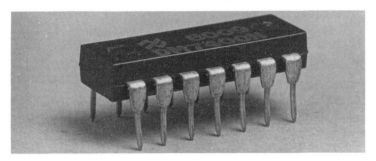

Fig.7.3 A typical IC.

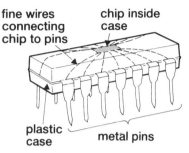

fine wires connecting chip to pins

chip inside case

plastic case

metal pins

Fig.7.4

Fig.7.4 is a drawing to show how each pin is connected to the integrated circuit inside the case. One of these pins is used to connect the IC to the positive supply rail and another to connect it to the negative supply rail. The remaining pins are used for input and output connections to whatever circuit or circuits the IC contains. The manufacturer tells you which pin is connected to what. ICs like these are very widely used and cost only a few pence.

As well as being small and cheap, ICs are also very reliable and will continue to do the job for which they were made without trouble for a long time. The photograph (Fig.7.5) shows the ICs in a microcomputer. A few years ago a computer containing this number of circuits would have occupied most of a room, it would have broken down often and would have cost hundreds of thousands of pounds!

Fig.7.5 Photograph of ICs inside a microcomputer.

59

Of course it is not just computers which contain integrated circuits. Nearly all pieces of electrical equipment – televisions, radios, washing machines, pocket calculators and so on – now contain 'the chip'. Its invention has revolutionised electronics.

Background reading

Changing capability and cost

The principal economic factor underlying the success of the new technology is the combination of increasing technical capacity and sophistication, with falling cost.

In 1950, Ferranti produced Europe's, and arguably the world's, first commercial computer, the Mark 1. The Mark 1 filled two bays around 5 metres long, 3 metres high, and 1 metre deep. It needed 27 kilowatts of power to function. Most of this power was converted into heat: air-conditioning was essential to avoid overheating. In present-day values, it cost around £1 million. In 1977, the same company produced the first European-developed microprocessor, the F100. The F10 was half a centimetre square and 0.6 millimetres thick. It had about 100 times the computing capacity of the Mark 1, operated on only 5 milliwatts of power, and sold for £50.

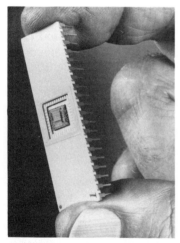

Fig.7.7 The Ferranti F100 microprocessor, which makes possible modern computers which are 10 million times smaller, 10 thousand times cheaper, and 100 times more powerful than the Ferranti computers built in the 1950s. This photograph was taken in 1977. Since then, microprocessors have reduced in size even more.

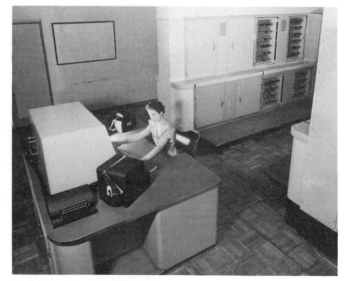

Fig.7.6 The Ferranti Mark 1 computer at Manchester University; the cabinets and operating console fill a large room. This photograph was taken in 1952.

Over the last 20 years the cost of computer hardware has fallen at a rate of around 30 % each year. Throughout the seventies the rate was nearer to 40 %. Even when inflation is taken into account, computer hardware costs about one thousand times less than it did 20 years ago. The computer industry expects to achieve a further reduction of the same size in the real costs of hardware over the next 20 years.

The standard, if slightly extravagant boast of the computer industry is as follows. If the motor industry had developed with the speed and efficiency of the computer industry, a Rolls Royce would now cost a few pounds, and do several million miles to the gallon. You would also be able to park five of them on your fingertip.

(From *Society and the New Technology* by Kenneth Ruthren.)

Background reading

The integrated circuit and the chip

In 1952, G.W.A. Dummer of the Royal Radar Establishment in Great Britain put forward the idea of an integrated circuit, but his ideas were not followed up. The effective integrated circuit was to be an American achievement. Engineers in several companies developed ideas for having a number of transistors in a single package and Jack Kilby of Texas Instruments Incorporated made the first working circuit in 1958. But, a new company, Fairchild Semiconductors, set out in 1957 to develop a novel concept in miniaturisation; namely the process in which transistors were created in the surface of a wafer of silicon. Success came by 1960. The process set scientists and engineers on a hectic race towards previously unimagined degrees of miniaturisation and complexity in microelectronic circuits.

Integrated circuits were first used in defence equipment and space vehicles but, by 1963, they were being used in commercial products, at first in small personal things like hearing aids, but soon afterwards in computers. These, in defence and space equipment then in business and industrial equipment, were to be the greatest users of integrated circuits.

A clear aim was to increase the number of transistors on a chip so that more computing capacity could be built into a given space. Fortunately the achievement of this also made for cheapness and, by eliminating wiring, reliability.

Fig.7.8 The 1966 Ferranti Argos 400 computer, one of the first to use integrated circuits. It is a desk-mounted computer, and folds into a box 30 cm × 20 cm × 30 cm.

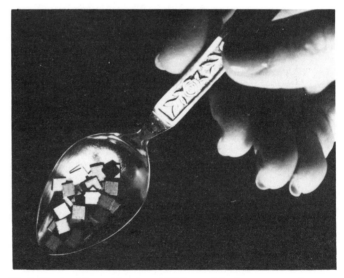

Fig.7.9 Each of these chips, from IBM's small business system, the System/36, is equivalent to the processing power of a 1960s computer.

In 1965, 30 components could be put on a 5 millimetre square silicon chip: by 1975, 30 000; by 1978, 135 000. But numbers have not been the only gain; the minute circuits are far more complex. The computer which filled a large room thirty years ago is now on a 5 millimetre chip.

(From *The Challenge of the Chip*, HMSO.)

Chapter 8 The quad NAND integrated circuit

Some things in this chapter – and also in Chapter 11 – will already have been mentioned in Chapter 6. They are repeated here because some readers may have omitted Chapter 6, which was about the relay version of the NAND gate, and gone straight to Chapter 7. If you did work through Chapter 6, you will now meet a more convenient form of NAND gate and you should find some parts of the chapter useful revision of ideas about NAND gates which you met earlier.

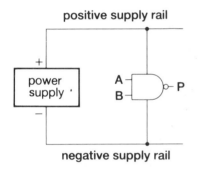

Fig.8.1

The NAND circuit is a switching circuit which is one of the basic building bricks from which more complicated circuits are made. It needs a power supply to provide a positive supply rail (high voltage level) and a negative supply rail (low voltage level). The NAND circuit is connected between these levels, as shown in Fig.8.1. This diagram includes the usual symbol for a NAND circuit.

The NAND circuit has two inputs, A and B, and an output P. The voltage level at the output P depends on the voltage levels at the inputs A and B. A voltmeter could be used to show the voltage level at the output, but so many voltmeters would be needed in the experiments which follow that it is more convenient to use LEDs (see Fig.8.2).

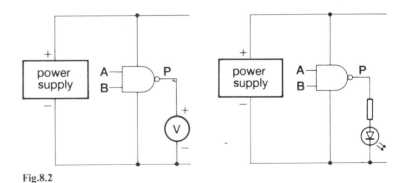

Fig.8.2

The experiments in this chapter use the quad NAND module (Fig.8.3), which has an IC containing four NAND circuits.

Note that the connections between the power supply rails and the NAND circuits are not marked on the module. In fact the positive and negative supply rails are connected to pins on the integrated circuit, and inside the chip these rails are connected to each NAND circuit.

Fig.8.3 Quad NAND module.

So all you have to do is connect a power supply to the *module* (Fig.8.4); the connections to the separate NAND circuits have been made for you.

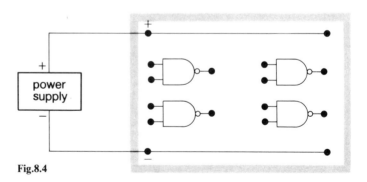

Fig.8.4

The module uses a single IC (or chip) containing four NAND circuits, as shown in Fig.8.5.

to positive supply rail

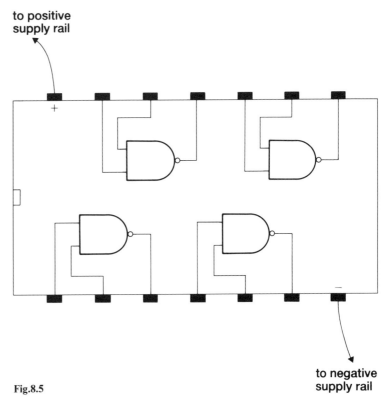

to negative supply rail

Fig.8.5

In circuit diagrams it is usual to show the symbol for a NAND circuit without the power rails connected to it. This, of course, saves space. But you should remember that a power supply must always be connected to the IC to make it work.

You will soon have the chance to explore the behaviour of the quad NAND module. But to do this you need a special LED indicator module (Fig.8.6). It has four LEDs on it and, like the quad NAND module, needs to be connected to a power supply (Fig.8.7). In the first experiment below you will find out how the LED indicator module behaves.

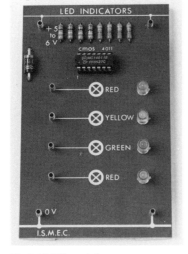

Fig.8.6 LED module.

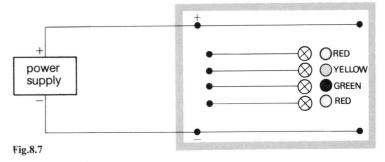

Fig.8.7

65

Experiment 8.1   The LED indicators

1. Set up the circuit shown in Fig.8.8 using the LED indicator module and a suitable power supply. Take care to connect the power supply correctly so that the positive side is connected to the positive power supply rail and the negative side to the negative power supply rail.

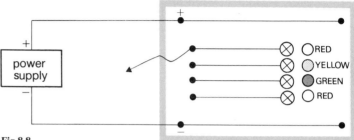

Fig.8.8

2. Now plug a lead into the top input socket of the indicator module. The other end of this lead may be connected to the positive power supply rail or to the negative power supply rail. Such a lead is called a flying lead. If connected to the *positive* rail, the input is said to be *high*. If connected to the *negative* rail, it is said to be *low*.

3. Find out how the top LED behaves when its input is connected first high and then low.

4. Do the other LEDs behave in the same way?

The experiment shows that each LED lights when the voltage level at its input socket is high.

Experiment 8.2   The NAND circuit

Set up the circuit shown in Fig.8.9 using one of the four NAND circuits on the quad NAND module, and one of the LED indicators. Make sure that they have their power supply rails linked correctly.

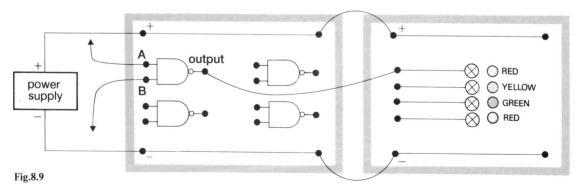

Fig.8.9

Each of the inputs A and B can be connected to the positive power supply rail (high) or to the negative power supply rail (low) using a flying lead. The output is *high* if the LED is lit and *low* if the LED is not lit.

1. By connecting the inputs high and low, complete the table:

B	A	Output
low	low	
low	high	
high	low	
high	high	

2. Does an *unconnected* input behave as if it is high or low?

Truth table for the NAND circuit

The table of results for the NAND circuit is shown below.

B	A	Output
low	low	high
low	high	high
high	low	high
high	high	low

A table of this kind is called a truth table. Notice that the output is:

Not high only when input A **AND** input B are high.

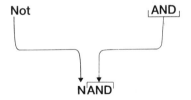

This is the reason for it being called a NAND circuit, though it is also frequently called a NAND *gate*. We shall see why it is called a gate in Experiment 8.3.

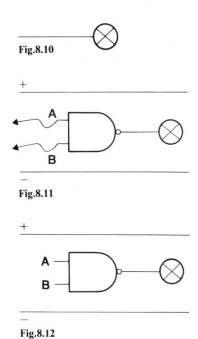

Fig.8.10

Fig.8.11

Fig.8.12

Circuit diagrams

In the circuit diagrams from now onwards, an indicator will be represented by the symbol shown in Fig.8.10.

To keep circuit diagrams simple, the power supply may not be drawn in future and only the parts of a module actually in use will be shown. The diagram of Fig.8.9 then simplifies to Fig.8.11. The + and − signs near the power rails show that a power supply is connected between them even though it has not been drawn.

Unconnected inputs

In question 2 in Experiment 8.2 on the previous page you were asked to find out whether an *unconnected* input to the NAND circuit behaved as if it were low or high. You should have found that such an input behaves as though it were *high*.

This means that, in a circuit such as Fig.8.12, the indicator will normally be off. (Look at the truth table for the NAND circuit: if A and B are both high, the output will be low.) It is possible to build NAND circuits such that unconnected inputs are low, but they are not used in the work described in this book.

Experiment 8.3   Using the NAND circuit as a gate

Add a push-button switch to the circuit of Fig.8.9 so that it is the same as Fig.8.13. Note that a simplified circuit diagram is again used.

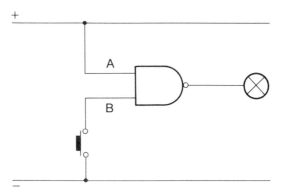

Fig.8.13

1. Press and release the switch several times regularly. What happens to the LED?

2. Transfer the connection at the other input from the positive rail (high) to the negative rail (low). Again press and release the switch regularly. What happens to the LED this time?

When one input, in this case A, is high, changes in the voltage level at B cause the LED to go on and then off, but when A is low the changes in the level at B do not have any effect at the output (check this from the truth table). This can be likened to a gate. When A is high, the gate is open and changes in the voltage level at B cause changes in the level at the output. But when A is low, changes in the level at B do *not* change the level at the output; it is just as though the gate had been closed.

Experiment 8.4 The NAND gate as an inverter or NOT circuit

Connect the two inputs of a NAND gate together as in Fig.8.14. You will notice that we are yet again using the simplified type of diagram which we mentioned earlier in this chapter.

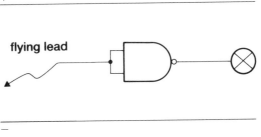

Fig.8.14

1. The two joined inputs can now be thought of as a single input. This single input can be taken high or low using the flying lead. By doing this, complete the truth table:

Input	Output
high	
low	

This experiment shows that the output is the opposite of the input and it is for this reason that it is called an *inverter*. Another name for it is a NOT circuit as it reverses the state of the input.

Experiment 8.5   Making an AND gate from two NAND gates

Connect the circuit in Fig.8.15 using two gates from the quad NAND module. Note that the second NAND gate is connected as an inverter (NOT circuit).

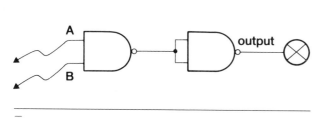

Fig.8.15

B	A	Output
low	low	
low	high	
high	low	
high	high	

1. The above circuit should behave as an AND gate. Check this by taking the inputs high and low, and completing the truth table.

The output of this circuit is only high when A **AND** B are high, and that is why it is called an AND circuit or AND gate.

Experiment 8.6   OR and NOR gates made from NAND gates

1. Draw and complete the truth table for an OR gate (output high if input A OR input B is high).

2. How does this compare with the truth table for a NAND gate?

3. Notice that if every high input to the NAND gate became a low, and every low input a high, it would give the OR gate you want. How can you change the inputs to the NAND gate? *Clue*: you have just learnt about inverters.

4. Now build an OR gate using three of the NAND gates on your module. When you have built the circuit, check that it produces the correct truth table.

5. Draw and complete the truth table for a NOR gate. Use the remaining NAND gate on the module to convert your OR gate to a NOR gate. Check that your circuit produces the correct truth table.

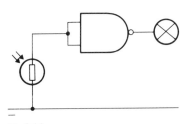

Fig.8.16

Experiment 8.7 Operating gates with LDRs

Connect an LDR between the input of an inverter and the negative supply rail, as in Fig.8.16.

1. What happens to the LED as the LDR is covered and uncovered?

In bright light the LDR behaves like a closed switch, so the input of the inverter will be low and the LED lit. In the dark the LDR has a very high resistance and behaves like an open switch. The input of the inverter is therefore unconnected and behaves as though it is high. This means that the LED will not be lit.

This is not really a very useful circuit – a light which comes on in the light and goes out in the dark! However, in Project 8(b) below you will have the chance to turn it into something better – a light which comes on automatically in the dark.

Experiment 8.8 Using a voltage divider

1. Use a potentiometer to set up the circuit in Fig.8.17. Turn the potentiometer spindle and see how the voltmeter reading changes.

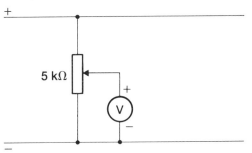

Fig.8.17

2. Set up the circuit in Fig.8.18 in which the NAND gate is connected as an inverter. Turn the potentiometer spindle so that the input voltage to the gate is zero as shown by the voltmeter. Is the LED on or off?

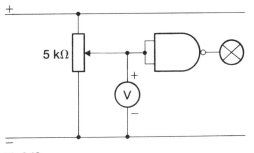

Fig.8.18

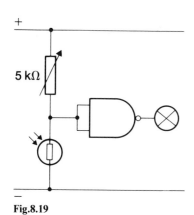

Fig.8.19

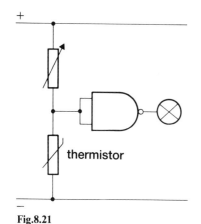

Fig.8.20

3. Now increase the voltage to the input and find out what happens.

4. What is the input voltage when the LED just goes off?

5. Now use an LDR and a variable resistor to set up the circuit of Fig.8.19.

6. With the LDR uncovered, adjust the variable resistor until the LED just comes on. Now move your hand towards the LDR to block off some of the light. What happens?

7. What must be the voltage across the LDR when the LED just goes off?

In the final experiment above, the variable resistor and the LDR are connected in series across the power supply. The voltage across them both must always add up to the power supply voltage. The power supply voltage is divided (shared out) between the two components: the variable resistor and the LDR act as a voltage divider (often called a potential divider).

Since the resistance of the LDR increases when it is shaded from the light, the voltage across it increases (the voltage across the variable resistor falls). The voltage across the LDR is also the input voltage to the inverter, so this voltage also increases. The LED is switched off when the switching voltage found in the first part of the experiment is reached.

In the experiments in this book you do not usually need to use a potential divider to switch the gates. The advantage of using a variable resistor and an LDR as a voltage divider is that the resistor can be adjusted so that switching takes place for only a small change in illumination. In other words, the circuit can be made more sensitive.

Experiment 8.9 Thermistors

A thermistor is a resistor whose resistance depends on temperature (see Fig.8.20). When heated, the resistance of a thermistor decreases. It can be used as one part of a voltage divider to make a temperature-dependent switch.

1. Set up the circuit in Fig.8.21. (The thermistor used should have a resistance of about 1000 ohms when cold.)

2. Adjust the variable resistor until the LED is just off. Now warm the thermistor by holding it tightly between your fingers. What happens to the LED?

Fig.8.21

Applications using NAND gates

One of the things which we are most anxious that you should learn is that electronic circuits can do useful things. You now have enough knowledge to design some circuits. We list below some projects. You will not have time to do them all, but we hope you will be able to solve some of the problems. Some of the projects are simple, others are quite difficult and you may need some help from your teacher.

Project 8(a) Burglar alarm (NAND application)
Use a single NAND gate, an LDR, a push-button switch and a buzzer to make a simple burglar alarm. The alarm should sound when the LDR is illuminated (the burglar's torch) or the switch is closed (the burglar's foot).

Project 8(b) Automatic night light
Use two NAND gates (both connected as inverters), an LDR and an LED indicator to make a circuit in which a light (the LED) will come on automatically in the dark. A circuit of this kind would be useful to light up the display in a shop window automatically when darkness falls. (*Hint*: look back to Experiment 8.7.)

Project 8(c) Length detector (AND application)
Use two NAND gates (one connected as an inverter) together with two LDRs and a buzzer to construct a system which will sound an alarm when an object is longer than a specified maximum length. Such a circuit might be used to reject overlong objects passing along a conveyor belt in a factory. (*Hint*: you can use an illuminated LDR to hold the input of a gate low.)

Project 8(d) Fire alarm
Use a NAND gate connected as an inverter, a thermistor, a variable resistor and a buzzer to make a simple fire alarm. The alarm should sound when the thermistor is heated. How would you adjust the circuit so that the alarm sounded when the temperature reached a higher value?

Project 8(e) Safety circuit for a safe
A safety circuit for a large safe sounds an alarm only if the door is closed but not locked. Closing the door *opens* a switch, whilst locking the door *closes* another switch. What sort of logic gate is required to do this? Set up the circuit and test it. Now adapt your circuit for a safe which has two locks.

Project 8(f) The skittle alley winner-indicator
In a fairground skittle alley, customers try to bowl over three skittles. Each skittle stands on a small switch which it keeps closed until it is bowled over. Design a circuit which lights a lamp only when a customer is successful.

Project 8(g) Car doors warning light

Design a circuit which will cause a warning lamp on the dashboard of a two-door car to light if either of the doors is not closed. Closing a door closes a switch.

Project 8(h) Circuit investigation

Fig.8.22 shows a circuit to be investigated, rather than a problem to solve. Copy the truth table and complete it to show the voltage level at each of the lettered points for each of the combinations at A and B.

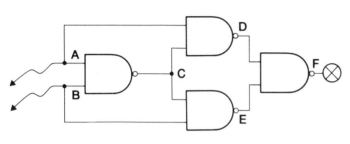

B	A	C	D	E	F
0	0				
0	1				
1	0				
1	1				

Fig.8.22

Then set up the circuit and check your predictions for the output F. If you have made an error, use a flying lead from another LED indicator to check your predictions for the level at each of the points C, D and E. Do this by touching the free end of the flying lead on to each of the points in turn.

Finally, suggest a use for this circuit.

Background reading

Getting the most out of chips

Putting more capacity into a smaller space is of enormous help to the advanced control systems employed in defence systems, space travel and complex industrial processes. And putting more components on to a chip clearly aids this process. But it does more. Wiring is potentially unreliable. Consequently the more wiring we can get rid of, the more likely are we to be able to improve reliability. And while the reliability of equipment is of considerable importance in military systems it is of growing importance to ourselves.

Reliability may also be increased because enormous numbers of electronic components are employed in modern systems and variations of performance between individual components are more probable than between integrated components. So the more functions we can fit into a chip the more reliable will they be.

It is clear that integration reduces enormously the cost of making up a circuit owing to the reduction in the amount of assembling and wiring involved. Moreover, due to the methods of manufacture employed, the cost of producing a transistor on a chip is very much less than that of producing an individual transistor. And this, of course, applies to other devices like resistors and capacitors. Finally, because so much more electronic capacity is being built into a much smaller space, the cost of making mounting systems, whether they are printed circuit boards or other arrangements, and the cost of casings and enclosures are also reduced.

Indeed it is the low cost of the chip as much as its capabilities that enables us to employ it in situations undreamt of even ten years ago.

So we have moved from the small scale integration of the early 1960s, amounting perhaps to 16 components on a chip, through medium scale integration to the large scale integration of today. Very large scale integration is about to begin.

(From *The Challenge of the Chip*, HMSO.)

Chapter 9 Logic circuits and truth tables – a summary

We have seen that some simplification of circuit diagrams can be made by leaving out the power supply. Later we shall leave out the power rails, though we must always remember that *all* modules must be connected between the same power rails.

More space could be saved by using one symbol to represent those circuits using two or more NAND gates. The symbol usually used for an AND circuit is shown below in Fig.9.1, together with its truth table. The symbol is not hard to remember – it is like a D and AND ends with a D.

AND

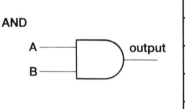

Fig.9.1

B	A	Output
low	low	low
low	high	low
high	low	low
high	high	high

This symbol can be used whenever it is not necessary to show that the AND circuit was made from two NAND circuits. But if we wish to show the two NAND circuits, we shall need to use the NAND gate symbol. The symbol for a NAND circuit is shown in Fig.9.2, together with its truth table. The only difference from the symbol for the AND circuit is the small circle at the output. The small circle simply means that the AND output has been inverted. You can see that by comparing the two truth tables.

NAND

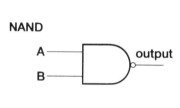

Fig.9.2

B	A	Output
low	low	high
low	high	high
high	low	high
high	high	low

How should an inverter be drawn? An inverter is a NAND gate with only one input or with the two inputs joined together, so it could be drawn as shown in either of the ways in Fig.9.3(a) or (b). Another common symbol for an inverter is shown in Fig.9.3(c).

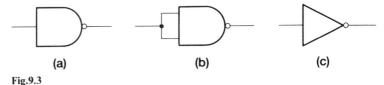

(a)   (b)   (c)

Fig.9.3

Because a NAND circuit with an inverter is the same as an AND circuit, diagrams (a) and (b) in Fig.9.4 are equivalent.

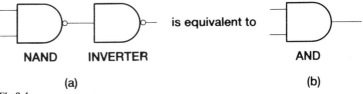

NAND INVERTER **is equivalent to**   AND

(a)   (b)

Fig.9.4

Circuits like the AND, NAND and inverter are used a great deal in electronics. They are called *logic circuits*. They all have inputs and outputs. They are all switching circuits in which the voltage level of the output at any time depends on the voltage levels at the inputs. They are 'decision-making' circuits.

In addition to those logic circuits mentioned above, there are two more which are useful. These are the OR gate and the NOR gate.

The output of an OR gate is high when either its A input **OR** its B input is high (or when both are high). The circuit symbol for an OR gate, together with its truth table, is shown in Fig.9.5.

B	A	Output
low	low	low
low	high	high
high	low	high
high	high	high

OR

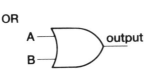

Fig.9.5

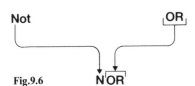

Fig.9.6

The output of a NOR circuit is Not high when either its A input **OR** its B input is high (Fig.9.6).

The circuit symbol for it is the symbol for the OR circuit with a little circle at the output. The circle shows that its output is the inverse of an OR circuit output (Fig.9.7).

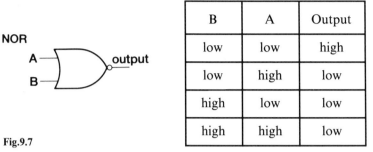

NOR

B	A	Output
low	low	high
low	high	low
high	low	low
high	high	low

Fig.9.7

It follows that an OR and an inverter are equivalent to a NOR circuit, and a NOR and an inverter are equivalent to an OR circuit, as shown in Fig.9.8.

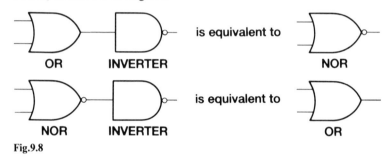

Fig.9.8

Instead of writing 'high' and 'low' in truth tables, it is more usual to adopt the convenient shorthand of writing 1 for 'high' and 0 for 'low'. Truth tables using this shorthand are on the next page.

Combinations of circuits

Circuits are often used in combinations, as shown in the example in Fig.9.9. To help you understand how this circuit works, we have put the type of circuit inside each symbol, though that is not usual with these diagrams.

To work out a truth table for this arrangement, it is necessary to build up a table which tells you what the voltage levels are at the intermediate points P and Q.

P is the output of an inverter, so that when input A is at a low level (0), P is at a high level (1), and vice versa. Copy the table on the next page and complete the P column.

Next, the Q column is the B column inverted.

Now that you know the P and Q voltage levels, you can fill in the output column because P and Q are connected to the inputs of the AND circuit.

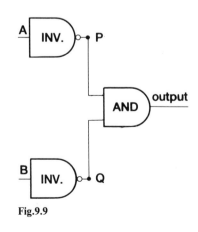

Fig.9.9

B	A	P	Q	Output
0	0			
0	1			
1	0			
1	1			

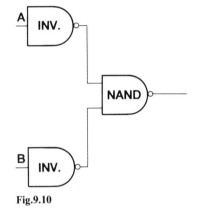

Fig.9.10

If you have filled in the table correctly, you will find that the combination is a NOR circuit.

What sort of circuit is the combination shown in Fig.9.10? You should find that it is an OR combination.

Summary

A summary of the logic circuits is given in Fig.9.11 below.

AND

B	A	Output
0	0	0
0	1	0
1	0	0
1	1	1

NAND

B	A	Output
0	0	1
0	1	1
1	0	1
1	1	0

OR

B	A	Output
0	0	0
0	1	1
1	0	1
1	1	1

NOR

B	A	Output
0	0	1
0	1	0
1	0	0
1	1	0

Fig.9.11

Background reading

Input–process–output

When you used a NAND module, you applied an input to it through its input terminals. Some process occurred inside the module. Then something happened at the output.

This is typical of modern electronics. It happens in computers: an input is processed into an output in a form convenient for use. This concept of input–process–output is not confined to electronics, as the following example shows.

Suppose you are running a dry-cleaning shop. The input consists of clothes delivered to the shop for cleaning. They go through the process of being cleaned, and then comes the output when they are packaged ready for the customer to collect. It is not necessary for the customers to understand the process: they are concerned only with the input and the output.

Of course a bit of organisation of the input may make the process more efficient – getting all the dresses together and all the suits – and the output can be organised in various ways.

The electronic modules you have used all have inputs and outputs, but the precise process inside does not really matter. Your NAND modules may have had reed relays, but other versions can be built using transistors – either separately or in an IC.

Chapter 10 **More questions**

1. Write out the truth tables (using 0 for low and 1 for high) for
   a an inverter,
   b a NAND circuit,
   c an AND circuit.

2. Write down the truth table for each of the following circuits which have their inputs connected together (Fig.10.1). Comment on your answers.

(a)   (b)   (c)   (d)

Fig.10.1

3. Logic circuits are often used in combination to form a system. In order to work out a truth table for the system, you need to work out what the voltage level is at each intermediate connection for the various inputs. For example, if you want to know what happens at the output of the system in Fig.10.2 for different input levels at A and B, you must first find out what the voltage levels are at P and Q for the various inputs.

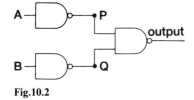

Fig.10.2

B	A	P	Q	Output
0	0			
0	1			
1	0			
1	1			

Copy the table and complete it. What sort of circuit is this combination?

4. Complete the following statements:
   a If both the inputs of a NAND circuit are left unconnected, the output is . . .
   b If only one input of a NAND circuit is used, and if that input is connected to the . . . supply rail, the output is . . .
   c In a NAND circuit the output is . . . only when . . .
   d In an AND circuit the output is . . . only when . . .
   e An OR circuit is so called because the output is high when . . .

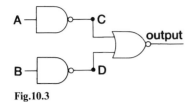

Fig.10.3

5. Copy the truth table for the circuit in Fig.10.3.

Input A	Input B	C	D	Output
0	0			
0	1			
1	0			
1	1			

a In the column headed C, put down the 0 or 1 for the output point marked C depending on what is written in the column headed A.

b In the column D put 0 or 1 depending on what is in column B.

c Use columns C and D to decide what should be in the output column.

d What sort of circuit is this?

6. Work out a truth table for each of the combinations in Fig.10.4.

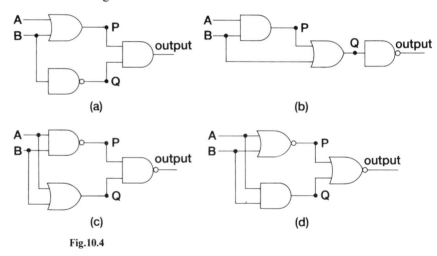

Fig.10.4

7. What is the minimum number of 2-input NAND gates needed to make the circuits of Figs 10.3 and 10.4?

8. Work out a truth table for each of the combinations in Fig.10.5.

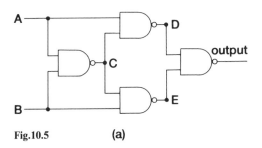

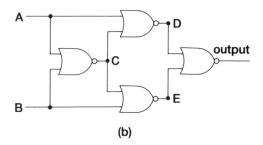

Fig.10.5   **(a)**   **(b)**

9. Fig.10.4(d) is similar to Fig.10.4(c), except that NOR gates have replaced NAND gates and an AND gate has replaced an OR gate. Compare the truth tables for the pair of circuits. What conclusion can you reach about the effect of these changes? Is the same conclusion true of the circuits of Fig.10.5?

10. You have two push-button switches A and B, a red LED, a green LED and a buzzer. Design a system, using logic circuits, which will cause the red LED to light if only A is pressed, the green to light if only B is pressed, and the buzzer to sound without the LEDs being lit if A and B are pressed together.

11. A thermistor is a special resistor whose resistance becomes smaller as it gets hotter. Draw circuit diagrams to show how it could be used to make a fire alarm using

  a   a battery and a buzzer alone;

  b   a battery, a buzzer and a reed relay;

  c   a battery, a buzzer and a NAND circuit.

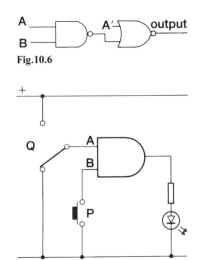

Fig.10.6

12. In Fig.10.6, inputs A and A′ are left floating (not connected to anything). What will the output be if (a) input B is low, (b) input B is high? Is this circuit any use?

How would its behaviour differ if input A′ were connected to the negative supply rail? Would it be any use then?

13. A student wishes to send out a message in Morse using the circuit in Fig.10.7. The idea is to flash the LED by means of the switch P, but it does not work.

  a   Explain why it does not work.

  b   What should be done in order to send out the message?

  c   When another student sees the circuit, she remarks that the message 'passing through' the circuit is just like passing through a gate, with Q acting as a lock. What do you think she meant?

Fig.10.7

Chapter 11 **The bistable circuit**

In this chapter, you will build a bistable using two NAND gates on the quad NAND module. This circuit is used on very many occasions in electronics and especially in computers. After exploring how the circuit behaves, there will be the opportunity to put it to some use.

Experiment 11.1   Making a bistable using two NAND gates

1. Set up the circuit shown in Fig.11.1 using two gates from the quad NAND gate module with an indicator on the LED indicator module connected to each of the outputs of gates 1 and 2. Do not forget to connect the power supply rails of *both* modules together.

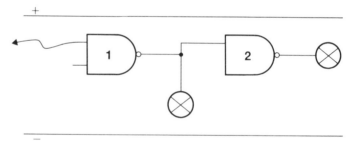

Fig.11.1

Notice that in Fig.11.1, the LED indicator connected to the output of NAND gate 1 is not drawn beneath the LED indicator of gate 2, even though that is where it is placed on the indicator module. Circuit diagrams are easier to understand if the parts are drawn in convenient places rather than as they are placed on the modules. Notice too that the power rails are drawn in the above diagram even though no leads are shown connected to them. Frequently, such power rails are omitted from circuit diagrams, but you must always remember that a power supply is needed to make the integrated circuits work.

2. Connect the flying lead to the low voltage level and note the voltage levels at the output of gate 1 and the output of gate 2.

3. Transfer the flying lead to the high voltage level and note the outputs of gate 1 and gate 2.

4. Explain what you observed in 2 and 3.

5. Now transfer the flying lead to the output of gate 2. Are the LEDs behaving as you would expect?

Up to now, both gates act as inverters. If the input to gate 1 is low, its output is high. As this output is connected to the input of gate 2, the output of that gate is low (Fig.11.2).

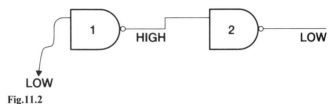

Fig.11.2

On the other hand, if the input to gate 1 is high, the output of gate 1 is low and so the output of gate 2 is high (Fig.11.3).

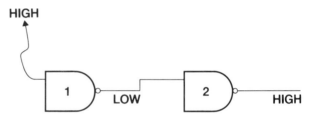

Fig.11.3

While you are transferring the flying lead which is attached to the input of gate 1 to the output of gate 2, the input to gate 1 remains at a high level because an unconnected input to a gate behaves as though it is high. When the connection is made, there is no change in the voltage levels in any part of the circuit because the output of gate 2 is also high (Fig.11.4). However long the connections are left, the LEDs will not change. It is a stable arrangement. But it is not the only possibility.

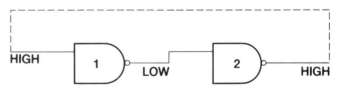

Fig.11.4

6. To obtain the other stable state arrangement, remove the two leads which connect the inputs and outputs of gate 1 and gate 2. Then connect the output of gate 2 to an input of gate 1, and, after that, connect the output of gate 1 to an input of gate 2. This should produce the other stable state of affairs. How does it differ from the first?

Fig.11.5 shows the other stable arrangement. What was at the high voltage level has changed to a low level, and what was low is now high. Because the circuit has two stable states, it is called a bistable circuit.

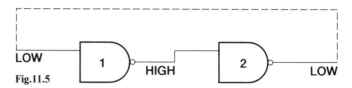

LOW

Fig.11.5

HIGH

LOW

For a circuit like this to be useful we must have a way of switching from one stable state to the other. The next experiment shows how this can be done.

Experiment 11.2 Switching a bistable circuit

Set up the circuit shown in Fig.11.6 with two push-button switches connected between the negative supply rail and the 'unused' inputs of gate 1 and gate 2.

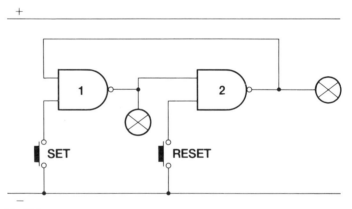

SET

RESET

Fig.11.6

1. First of all, press and release the switch marked RESET. Which LED is on and which one off?

2. Now press and release the switch marked SET. What happens?

3. Press and release the SET switch several more times. What happens?

4. Now press and release RESET. What happens?

5. Finally press SET and RESET repeatedly, one after the other. What happens?

Pressing a switch causes the bistable to switch from one state to the other *provided* it is the switch that is connected to the NAND gate whose output is low (that is, the one with the LED not lit).

To explain this, consider the bistable circuit in the state shown in Fig.11.7. If the SET switch is pressed, one input of NAND gate 1 is connected to the low voltage level, and so its output goes high. The output of gate 2 therefore goes low because both its inputs are high and in those circumstances the output of a NAND gate is low.

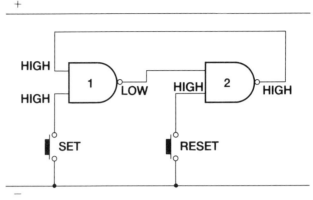

Fig.11.7

This low level is fed back to an input of gate 1, which keeps the output of gate 1 at a high level even when the SET switch is released. Pressing and releasing the SET switch has no further effect, since making the inputs to gate 1 both low, or one low and one high, does not change the output of gate 1 from high. The situation is now as in Fig.11.8.

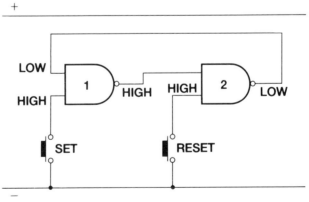

Fig.11.8

The system can only be switched back to its original state by pressing the RESET switch. This takes an input of gate 2 low so that its output goes high. Gate 1 now has two high inputs (remember an unconnected input floats high) and hence its output goes low. And this low input to gate 2 prevents any further change when the RESET switch is released.

Note that the switches must be connected between the inputs

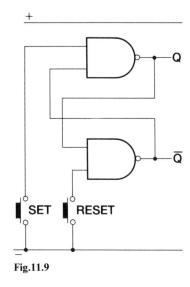

+

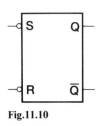

SET   RESET

Fig.11.9

S   Q

R   Q̄

Fig.11.10

and the negative supply rail. If they are connected between the inputs and the positive supply rail, they will have no effect at all, for then the input is high whether the switch is pressed or not.

An alternative way of drawing a bistable circuit using two NAND gates is shown in Fig.11.9. Q represents the output of the NAND gate to which the SET switch is connected. Pressing the SET switch *sets* Q to the high voltage level or 1. Q̄ represents the output of the gate to which the RESET switch is connected. Pressing RESET *resets* Q to the low level or 0.

Circuit diagrams again

If it is not necessary to show the two NAND gates, then it is more common to use the symbol shown in Fig.11.10. S and R stand for SET and RESET. Q and Q̄ represent the two outputs, as above. They are often called 'complementary' outputs, because when one is high, the other is normally low.

The small circles on the S and R input lines indicate that, in order to change from one stable state to the other, one or other of the inputs S and R must be taken low.

The bistable circuit is another very useful building block when it comes to designing circuits. It is often known as an 'RS bistable' or an 'RS flip-flop'. We can summarise how it works as follows.

(a) There are two stable states:
    the SET state when Q = 1 and Q̄ = 0
    the RESET state when Q = 0 and Q̄ = 1.

(b) Normally, the circuit will be in one of these stable states.

(c) To change from the SET state to the RESET state, the R input must be taken low. If R goes high and is taken low again, no further change occurs.

(d) To change from the RESET state to the SET state, the S input must be taken low. If S then goes high and is taken low again, no further change occurs.

Applications using the bistable circuit

The fun of electronics is in doing things with it; so once again we have a number of projects to test your ingenuity.

Project 11(a) Latched burglar alarm

Use two NAND gates, two push-button switches and a buzzer to make a latched burglar alarm. One switch should correspond to the "trip" switch. If this is closed (perhaps by the burglar's foot) the alarm should sound and stay on even when the switch is released or pressed again. The second switch should correspond to the "reset" switch, and would be hidden away in a place known only to the householder. Only when this switch is pressed should the alarm be silenced.

Once you have built this circuit, convert it to an alarm which will come on and stay on when a light (the burglar's torch) shines on an LDR.

Project 11(b) Latched fire alarm using a bistable

Build a fire alarm which, once triggered, will continue to sound until it is reset.

Project 11(c) Simple stop–go traffic lights

Traffic in a one-way street has to cross a bridge which can only have one car on it at a time. The bridge is to be controlled by a set of stop–go lights, activated by switches in the road, so that the lights go red when a car enters the bridge and then green as the car leaves it. Design a circuit to do this job.

Project 11(d) Traffic lights operated by an SPST switch

A simple set of stop–go traffic lights is to be operated by an SPST switch so that either the red light is on or the green light is on, but not both together. Design a circuit, using a bistable, which will do this.

Project 11(e) Quiz master

A quiz master circuit is used to identify the first contestant to push his or her answer button in a quiz game. Each contestant has a push-button and an indicator light (in this case, an LED). The light of the first person to answer should come on and stay on. At the same time all other lights should be prevented from coming on.

To make such a quiz master is not easy, and you may need help from your teacher. Each contestant in the game will need four NAND gates (two connected as a bistable), a push-button switch and an LED indicator. The best approach is to wire up the four gates for one contestant, and then wire up the four gates for another contestant separately. Finally connect the two sets together and see if your quiz master works!

Background reading

Using microelectronics today

The emergence of microelectronics has been so swift that there may seem to have been little time to put it into practice. Yet there are few areas in our lives which microelectronics has not affected. A major reason for this is that microelectronic products are cheap and easily built into a wide variety of systems by economical production tech-

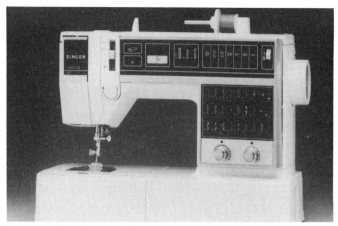
Fig.11.11 Singer 'Symphonie 300' sewing machine, with microprocessor-based switch selector.

niques. Another reason is that many devices employing microelectronics can be developed without spending large amounts of money and time.

The home

Whether we regard digital watches and pocket calculators as prestigious toys or necessary aids, these products are now universally available in wide variety, and we buy them avidly as we buy the many toys and games which are now appearing.

In addition, microelectronics is providing better control systems for washing, cooking and heating equipment. For some time we have had pre-set washing machines and pre-set cooker controls. But they have been bulky and expensive to make by comparison with the microelectronic systems which are now taking over. The latter are cheaper, provide more programs and are more reliable.

At present, cookers, washing machines and dishwashers are being fitted with their own microelectronic programmers. But the standard microprocessor and memories which are already available could enable us to program all the services we use in the entire house, for example to regulate the heating system in the most efficient manner, switch on lights, draw curtains, switch on the television set for selected programmes, record them, record telephone messages and produce automatic reminders of appointments, all from one central control unit. Is all this absurdly luxurious? Barely a hundred years ago, switching on electric light would have seemed a dream to most people.

(From *The Challenge of the Chip*, HMSO.)

Chapter 12 **Drivers**

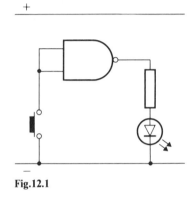

+

Fig.12.1

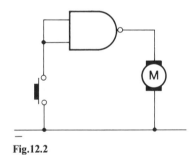

+

Fig.12.2

Experiment 12.1   Using a NAND gate to drive a motor

1. Set up the circuit shown in Fig.12.1, using a single LED indicator to show the voltage level of the NAND gate output. The NAND gate is connected as an inverter. Check that the LED is lit when the push-button switch is pressed.

2. Replace the LED with the motor module, as in Fig.12.2. What should happen when the push-button switch is pressed? Try it.

You might have expected the motor to turn when the switch was pressed, but it does not do so. The reason for this is that the IC logic gate only operates correctly when small currents, no greater than a few milliamperes, are involved. And your motor requires more than a hundred milliamperes to run it!

3. Connect the LED in parallel with the motor as in Fig.12.3. What happens when the switch is pressed?

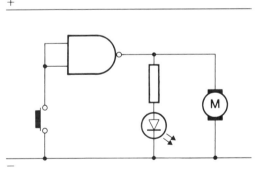

+

Fig.12.3

You will find that the LED does not light this time as it did in **1** above. This shows that the output of the NAND gate did not go high, but remained low. Any attempt to get a large current to flow from the output of a NAND gate prevents the output from becoming high.

Of course it is important to overcome this difficulty. This is done by introducing a *driver*. The driver is a circuit which will operate from a NAND gate without needing too much current, but will give enough current at the driver output to operate a relay switch. The driver/relay module is shown in Figs 12.4 and 12.5. The next experiment shows how the module is used.

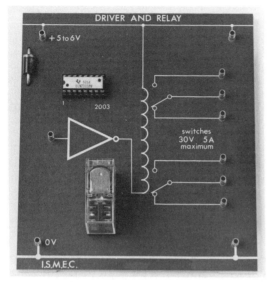

Fig.12.4 Driver/relay module.

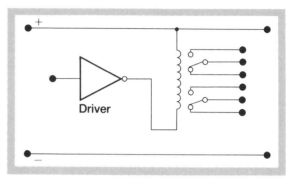

Fig.12.5

Experiment 12.2   Using the driver/relay module to switch a motor on or off

Connect the circuit of Fig.12.6 using the quad NAND and the driver/relay modules. Note that a separate supply is used for the motor and only one of the two SPDT switches operated by the relay is needed.

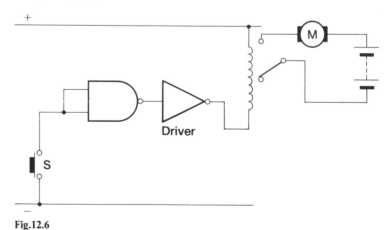

Fig.12.6

1. What happens when the switch S is pressed?

2. When the switch S is pressed, is the output of the NAND gate high or low?

3. The driver on the driver/relay module is a special kind of inverter. When the switch S is pressed, is the output of the driver circuit high or low?

4. The relay coil is connected between the positive supply rail and the output of the driver. When a current flows through the coil, the relay switch operates and the motor turns. When this happens is the output of the driver high or low?

5. Replace switch S by an LDR. What happens to the motor when the LDR is brightly lit and when it is dark? Can you think of a use for this circuit? Can you make the motor turn the other way?

A driver provides what is called an *interface* between an IC output (which can only control small currents) and a device such as a relay or motor (which requires relatively large currents).

Applications using the driver/relay module

Project 12(a) Reversing an electric motor
Using a push-button switch, a NAND gate and the driver/relay module, set up a circuit so that the direction of rotation of a motor is reversed when the switch is pressed. You will need both of the SPDT relay switches, with the motor connected between the moving contacts (see Experiment 3.5 on page 20). A separate power supply should be used for the motor if possible. Is the NAND gate really necessary?

Project 12(b) An automatic light
Use a driver/relay module, an LDR and any necessary gates to make a circuit in which a *filament* lamp will come on automatically in the dark.

Project 12(c) Reversing an electric motor with a bistable circuit
Using two push-button switches, two NAND gates and the driver/relay module, set up a circuit which reverses the direction of rotation of a motor when one of the switches is pressed and released. Both the SPDT relay switches will be required with the motor connected between the moving contacts (see Experiment 3.5 on page 20). A separate power supply should be used for the motor if possible. Explain how your circuit works.

Background reading

The new technology and women

The economic and social changes produced by the new technology may have important implications for the role of women in society. The trend towards home working and work sharing will make it easier to combine work with domestic responsibilities, child-rearing in particular. Not only will this improve women's opportunities to work; it will also make it possible to divide domestic responsibilities more fairly between men and women. And it will further undermine the distinction between paid work in the office or factory, and unpaid domestic work.

Moreover, as traditionally male unskilled jobs disappear with factory automation, and traditionally female 'caring' jobs in the service sector (nurses, teachers and social workers) grow, more and more men are likely to find themselves in what were previously thought of as 'female' occupations. At the same time, skill shortages may be reduced by making better use of women's abilities, and encouraging them to enter traditionally 'male', skilled occupations – for example as doctors, lawyers, financiers, engineers. These trends have, of course, already started. But their development, as a consequence of the introduction of the new technology, makes a further extension and equalisation of opportunity for women possible. In particular, as jobs start to lose their traditional 'male' and 'female' labels, the status and the financial rewards of traditionally female occupations may rise. Society may start to value caring more highly.

(From *Society and the New Technology*, by Kenneth Ruthren.)

Chapter 13 **Problem solving**

At the end of Chapter 6 and of Chapter 12, there was a project problem for you to solve. It was not a very hard one: you were told you had to make a motor turn round one way and then the other, and you saw how to do that in Experiment 2.6 on page 13. You know that if the motor is connected to the positive and negative supply rails as shown in diagram (a) in Fig.13.1, it will turn round one way. If it is connected as in diagram (b), it will turn the other way.

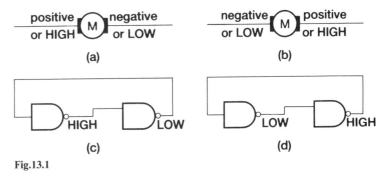

Fig.13.1

Next, you were told that you had to make these voltage changes using the two stable states of a bistable.

The two states of a bistable are shown in diagrams (c) and (d). Can you see how the motor should be connected for the bistable to control its direction of rotation?

The last step of all is to connect into the circuit the two push-button switches which will be used to cause the bistable to flip over from one stable state to the other. And you learnt how to do that in previous chapters.

Now you can draw the circuit diagram of the circuit which solves the problem. Set it up and try it if you did not succeed before (Fig.13.2).

To solve that problem you had to know how to work the motor (the output component), which circuit to use to do it, and how to use the switches (the input components) connected to the NAND modules to make the circuit work as you wanted it to do. Then you drew a circuit diagram, set it up and tested it. All the project problems can be solved in this way, but you will not always be told which circuit you should use. Of course, if the circuit does not

Using NAND relay circuits

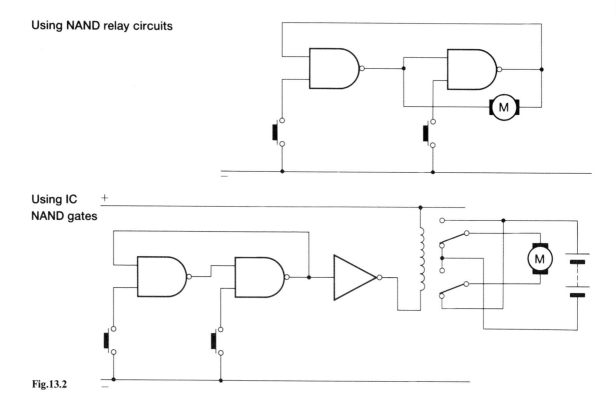

Using IC NAND gates

Fig.13.2

work correctly when you test it, then you have to think out why and put the error right.

Here is another project problem for you to solve, but this time you are not told anything about the circuit you should use.

Project   Vehicle moving backwards and forwards between light beams

A vehicle driven by an electric motor, such as a model car or train, is to move backwards and forwards between two beams of light (Fig.13.3).

Suppose the vehicle is moving to the left. When it interrupts the light beam on the left, the circuit must reverse the motor so that the vehicle moves to the right. And then, when it interrupts the right-hand light beam, the circuit must reverse the motor again.

It seems that this project is very similar to the previous one. There are two states for the vehicle: moving to the left and moving to the right. This suggests you should use a bistable circuit. And the bistable circuit needs to be switched from one stable state to the other by input components which are sensitive to light, instead of push-button switches. That suggests that LDRs should be used.

So you might think that one of the circuits shown in Fig.13.4 is the answer to the problem.

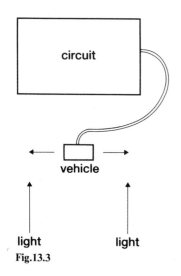

Fig.13.3

Using NAND relay circuits

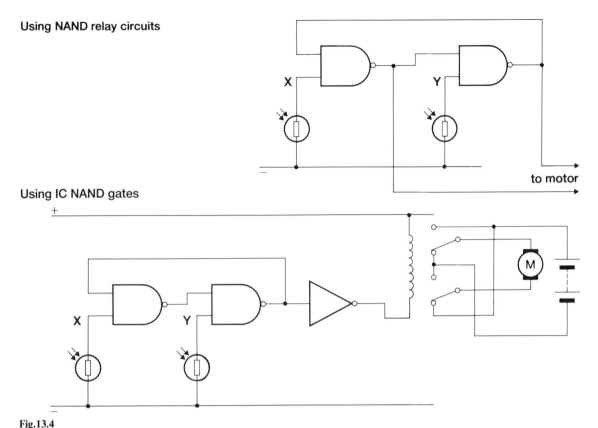

Using IC NAND gates

Fig.13.4

Would the circuit work? Think hard! What happens when the vehicle is between the light beams so that both LDRs are brightly illuminated?

If you still think it would work, set up the circuit and try it.

You will find that, when light falls on both LDRs, the circuit does not work. This is because the resistance of an LDR is small when bright light falls on it. That means that the inputs X and Y are connected to the negative supply rail and both are low. Both the outputs are therefore high and the bistable circuit is not in one of its stable states. If you used the top circuit of Fig.13.4, with both outputs high, no current will flow through the motor. If you used the one below, the motor will only turn one way when the vehicle is between the LDRs.

To get over this difficulty, what you now have to do is to find a way of making the inputs X and Y high (instead of low) when light is falling on the LDRs. Can you think how to do that? Which extra circuit do you need? Look back to Experiment 6.2 on page 43 or Experiment 8.4 on page 69.

If you use an inverter between each LDR and the input X or Y, then X and Y will be connected to the high level when the LDR is brightly lit. And when both LDRs are illuminated, the bistable will be in one of its stable states and the motor should turn correctly.

Now you should draw the correct circuit diagram and let your teacher see it. If it is correct, set up the circuit and test it.

You may find that it still does not work as you expect, even though the LDRs are brightly lit and the vehicle moves. Perhaps the vehicle moves right through one of the beams without the bistable switching to its other stable state. That could be because it has to be the *other* beam which needs to be interrupted to reverse the motor. How can that fault be put right? Try changing round the connections to the motor.

Solving other problems

Each project problem can be tackled in a similar way. First you must decide what is necessary to make the output components work, whether they are lamps, buzzers, motors or something else. Then you must decide which circuit to use to do that, and which input components are needed (switches, pressure pads, LDRs or something else).

Next draw a circuit diagram, check that it will work in the way you think it should, set it up and try it. If it does not do what you expect, try to work out why it is misbehaving. When you have done that, you should know how to put it right.

The following are suggestions for further projects.

Project 13(a) Window alarm
Design a circuit which will sound an alarm if a window is opened. You may imagine that a magnet is hidden in the window sash (the moving part). A reed switch is buried in the frame so that a magnet is near it when the window is closed. Once the alarm sounds, it should continue even if the window is closed again; you will need to have a switch to reset the alarm.

Project 13(b) Automatic signal for a railway
Use a bistable to make an automatic signal for a model railway. The train should turn the signal to red as it passes and then back to green when it has travelled some distance further. Two reed switch modules can be placed between the rails a suitable distance apart, and a small magnet can be taped underneath the last coach of the train.

Project 13(c) Fire alarm
Design a circuit which will sound an alarm if a fire occurs. Use a piece of fine copper wire or solder which would melt in a fire. When you test your circuit, cut the fine wire with scissors to pretend that it has melted.

Project 13(d) Light-controlled motor with emergency 'over-ride'
Build a circuit which will automatically turn off a motor when it gets

dark. Add an emergency 'over-ride' switch (use an SPDT module) which can be used to keep the motor going even though it is dark.

Project 13(e) Car doors warning light

Build a circuit for a two-door car which will light a warning indicator on the dash-board if both doors are not properly shut.

Use two push-button switch modules to act as switches mounted in the doors, and an LED module as the warning light. The light should be on except when both switches are pressed.

Project 13(f) Daytime sleeper's doorbell

Suppose you work nightshifts and sleep all day. Build a circuit which will stop your doorbell ringing in the day when you are asleep, but will let it ring when it gets dark.

Use a push-button switch to act as the doorbell button, a buzzer to act as the bell, and the LDR to sense if it is day or night.

When you have built your circuit, change it so that the doorbell only sounds in the daytime and not at night.

Project 13(g) Night-time anti-theft device for cars

Build a circuit which makes it impossible to operate the starter motor of a car at night.

Use the motor module to act as the starter motor, and a push-button switch to act as the ignition switch.

Project 13(h) Night-time only latched burglar alarm

In Experiment 6.7 on page 54 or Project 11(a) on page 88, you built a latched burglar alarm using the bistable circuit shown in Fig.13.5.

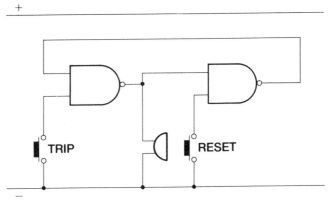

Fig.13.5

If the RESET switch is pressed, the buzzer is off. If the TRIP switch is pressed, the alarm sounds and stays on until the house-holder presses the RESET switch again.

The problem of this project is to build a latched alarm that will work only in the dark. In the daytime, operating the trip switch should have no effect. At night, it should sound the alarm.

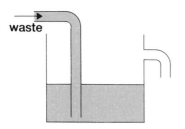

Fig.13.6

Project 13(i) Pollution warning system

A factory discharges waste liquid (which conducts electric current and which may be dirty) into a large tank with an overflow pipe (Fig.13.6). The waste is allowed to overflow into a river provided the liquid is clear.

Design a circuit which will sound a buzzer if the tank is full of dirty liquid.

Project 13(j) Seat-belt warning light

A car is to have a pressure switch inside the front passenger seat and a switch in the seat-belt mechanism which opens when the seat-belt has been fastened. Design a circuit which will show a red light if a passenger has not fastened the seat-belt, but a green light otherwise. (Remember that there may not be a passenger in the car.)

Background reading

New technology and the Third World

The main direct impact of the new technology on international relations is likely to be economic. In particular it may affect the relationship between the developed economies of North America, Japan and Europe, and the countries of the Third World.

The development of more technologically sophisticated products and the automation of manufacturing threaten the developing countries' share of the world market in manufactured goods. In recent years newly-industrialised countries such as Korea, Malaysia, Hong Kong, Singapore and Taiwan have captured a significant share of the world market in areas such as consumer electronics, textiles and heavy engineering. The main competitive advantage of these countries has been their low labour costs. These countries can produce radios or suits or oil tankers more cheaply because, although they do very similar jobs to Western workers, the workers who produce these goods are paid wages much lower than their Western counterparts. But industrial automation threatens to eliminate this advantage. For, by replacing expensive labour with machines, factories in the developed countries will be able to bring their production costs closer to those of their Far-Eastern competitors, and so take custom away from them.

(From *Society and the New Technology*, by Kenneth Ruthren.)

Chapter 14 Electronic control systems

In most of the experiments and projects in this book, electronic circuits have been used to switch motors, LEDs and buzzers on or off. In other words, circuits have been used to control these devices. When electronic circuits are connected together to control things, the whole arrangement is called a control *system*.

Systems are used, for example, to control what happens in aircraft, in power stations, in household appliances such as washing machines and microwave ovens, and for road traffic control and so on. Robots, operated by electronic control systems, are used nowadays, for example, in car factories. Electronic control systems are of ever-increasing importance in the world today, and the art of electronics lies in designing such systems.

This chapter looks at a simple control system in some detail to see how it might be built, first by using simple switches and then by using NAND circuits as the building bricks. There are not many experiments in it for you to do, but we hope you will read the chapter. It also tells you more about logic circuits and how their truth tables can help in the design of a system.

Machines which make decisions

Suppose your parents ask that you should make them a pot of tea each day at 4.00 p.m. – if they want it. Then each day you have to make a series of decisions. Is it 4.00 p.m.? Is tea wanted? Are a teapot and tea available? Is there a kettle with water in it?

Each of these questions can be answered by 'yes' or 'no'. If the answer to each of them is 'yes', then you would make the tea and tell your parents when it is ready. And since it would be a waste of

Fig.14.1

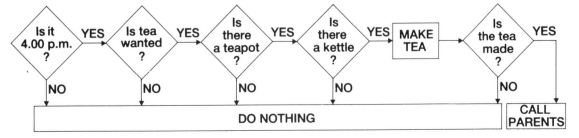

101

time to make tea if it were not wanted, you need to know the best order in which to make your decisions.

The series of decisions can be shown by a flow chart (Fig.14.1).

Of course you could make the decisions in another order as you can think for yourself. But suppose you decide to invent a machine to do this job for you. A machine cannot think for itself so the 'decisions' have to be in the right order. You may notice that each decision is like a switch. If the answer is 'yes', you go to the next question; if it is 'no', you do nothing. Here is an electrical equivalent of the flow chart (Fig.14.2).

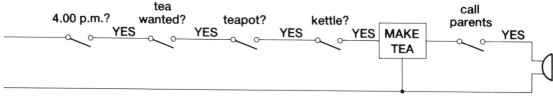

Fig.14.2

This electrical flow chart could form the basis of a machine. If an electricity supply were connected at one end, and if the first four answers were 'yes', the electricity could be connected to the heater in the kettle to boil the water.

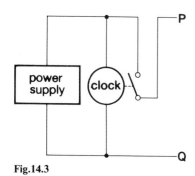

Fig.14.3

Three other parts are needed besides the switches: a time-switch, a buzzer and a special 'tea-maker'. The time switch is simply an electric clock which closes a switch at a set time, and keeps it closed for, say, half an hour. This is represented by Fig.14.3. (What would you have if you connected a buzzer between P and Q?) This time-switch will be the first switch in the electrical flow chart.

The most difficult part of the machine to invent is the part which makes the tea. The diagram (Fig.14.4) shows you how that can be done. The kettle-part has a tightly fitting lid with a tiny hole in it and an electric heater inside. When the water boils, the steam cannot escape quickly and the extra pressure, caused by the steam, pushes the boiling water out through the tube into the teapot.

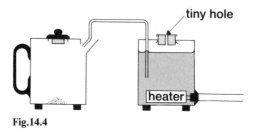

Fig.14.4

But the problem is not solved yet. The machine cannot *see* if the kettle has water in it or if the teapot is in the right place. This is why the decisions cannot be in any order. It would not do if the heater were switched on without the teapot in place or without water in the kettle!

To tell the machine whether the teapot is there and whether the kettle has water in it, we need a pressure switch for each of them. The teapot and the kettle are on small platforms standing on short springs. The platforms are pressed down when the teapot or the kettle full of water is in place. The springs supporting the kettle platform need to be stronger than those under the teapot platform because the kettle must not be able to press its platform down when it is empty. Each platform is fixed to an SPDT switch inside the base on which the platforms rest (Fig.14.5).

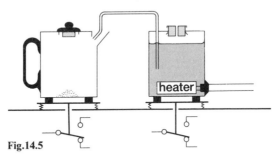

Fig.14.5

You may be worried about how the heater will be turned off when the tea has been made. If the electric current to the heater has to pass through the kettle switch, then, when the tea has been made and the kettle is nearly empty, the switch will change over and turn the heater off. And we can use the other contact of the switch to sound the alarm buzzer!

Now we can draw the circuit with its 'flow diagram' of switches (Fig.14.6).

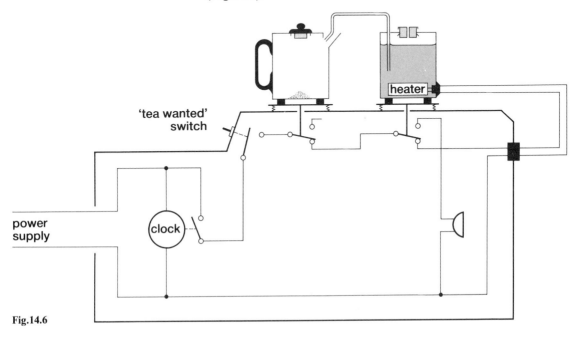

Fig.14.6

Using truth tables to design systems

What was described above was a *system* – a system of switches which allowed a useful job to be done. You may have noticed that the switches were in series, so forming an AND arrangement. That is to say, the electricity supply is only switched through to the heater if it is 4.00 p.m. AND tea is wanted AND the teapot is in place AND the kettle is there with water in it. You have found in previous chapters how to make an AND gate using two NAND gates. So it should be possible to design a system for the automatic tea-maker using NAND modules.

Truth tables can help to solve the problem of which circuit to use when you are designing a system. To show how this is done, take the problem of the 'daytime sleeper's door bell'.

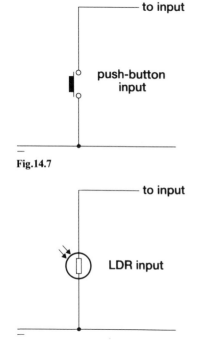

to input

push-button input

Fig.14.7

to input

LDR input

Fig.14.8

The daytime sleeper's doorbell

a. The circuit has to prevent the bell from ringing during the day.
b. A push-button switch is to act as the doorbell switch.
c. An LDR is to sense whether it is night or day.
d. A buzzer can play the part of the bell.

The push-button switch is an input component and has to be connected between an input terminal and the negative supply rail (Fig.14.7). When the switch is open, the input is not connected and it 'floats' high. When the switch is closed, the input is low.

The LDR is the other input component (Fig.14.8). It is connected between an input and the negative supply rail. At night, its resistance will be large and it will act like an open switch and the input will float high. In the daytime its resistance will be low and it will be like a closed switch, so that the input will be low.

You want the buzzer to sound only when the door switch is closed during the night. The buzzer will sound only when the output is high. We can begin to make up a truth table:

Door Switch	Time	Buzzer
open	day	silent
open	night	silent
closed	day	silent
closed	night	sounds

which can be written

Switch Input B	Time Input A	Output
1	0	0
1	1	0
0	0	0
0	1	1

If you look at the truth tables on page 79, the only ones with outputs which have three '0's and a '1' are the AND circuit and the NOR circuit, but the inputs are different. Let us look at the AND circuit.

The AND circuit truth table:

B	A	Output
0	0	0
0	1	0
1	0	0
1	1	1

The truth table needed:

B	A	Output
1	0	0
1	1	0
0	0	0
0	1	1

How are these different? How is the B column different?

If the B column of the AND truth table is inverted, we get the truth table needed.

So using an inverter on the B input (the door bell switch) will produce a system which will do the job required. The circuit is given in Fig.14.9.

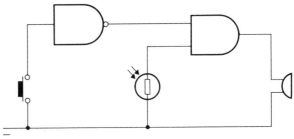

Fig.14.9

If only NAND circuits are used, the circuit is that in Fig.14.10.

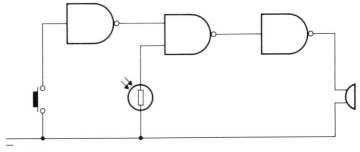

Fig.14.10

105

The automatic tea-maker again

It is possible to design the system for the automatic tea-maker using logic circuits. In that machine, the kettle heater was to be switched on only if each of the four switches were closed. Since a closed switch must connect an input to the low voltage level (remember that an open switch causes an input to float high), the truth table for what is needed is:

Time?	Tea?	Teapot?	Kettle?	Heater
0	0	0	0	1 (on)
Anything other than the above				0 (off)

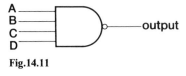

Fig.14.11

This time there are 4 inputs, and the circuits used for the experiments in this book have only 2 inputs! But there is nothing to prevent as many inputs as one wants being added to the NAND relay circuit, provided each has its own diode, and IC NAND gates can be bought which have 4 inputs (and even 8 or 13) – see Fig.14.11. And even if only 2-input NAND gates are available, a 4-input NAND gate can easily be built from them. See if you can show that the circuit of Fig.14.12 is a 4-input NAND gate. How many 2-input NAND gates would be needed to make it? The answer is five.

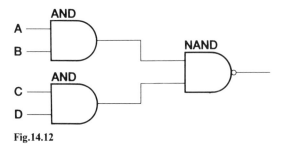

Fig.14.12

In the case of a 4-input NAND gate, only if every one of the inputs is high (or not connected) will the output be low. The truth table for this gate would be:

A	B	C	D	Output
1	1	1	1	0
Anything other than that				1

Now compare the table at the bottom of page 106 with the earlier one above it. All the inputs have to be inverted, and the output too. The output would have to switch the heater on by using a relay because the heater requires a lot of current. The circuit becomes that in Fig.14.13.

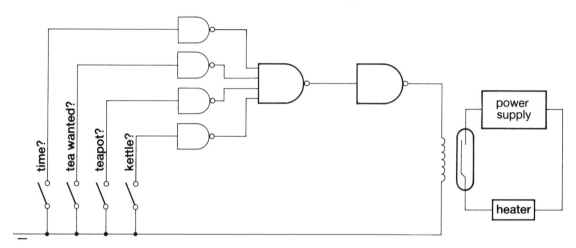

Fig.14.13

That leaves just the alarm buzzer to add. Remember that the buzzer has to sound when the tea has been made; that is to say, when the 'Time', 'Tea wanted' and 'Teapot' switches are closed, but the 'Kettle' switch is open. Another 4-input NAND circuit and an inverter are needed to operate the buzzer. See if you can do it!

This electronic system is more complicated than just four switches in series, but it has got some advantages. In the 'switches-only' circuit, the current for the heater from the heater supply has to pass through the switches. The switches have to be able to carry that current and be well enough insulated to protect the user from the dangers of high voltages. The NAND circuits work from *much* smaller voltages which are safe, and extra insulation is not needed. Indeed, except for the heater, the system could be operated from a battery.

But what about the expense of all the logic circuits? If the circuit 'building blocks' were NAND *relay* circuits, the system would be expensive without doubt. But if IC NAND gates are used, then all the logic circuits for the 'automatic tea-maker' system could be manufactured on a piece of silicon about the size of a pinhead! This is what is commonly called a 'chip' and the chip for this system would cost less than £1! No longer do you have to be a millionaire to own a computer!

That is why electronics has become so important. Very complicated circuits can be made very small and very cheaply.

Background reading

The new technology and the disabled

One group which will undoubtedly benefit from the new technology is the disabled. Microelectronics promises to give many disabled people an effective means of living a normal life.

Take the example of one man, handicapped for 20 years by multiple sclerosis. Now, thanks to the new technology, he is able to earn his living by running an accountancy business from his home. He instructs his microcomputer by breathing through a suck–blow tube, as he is unable to use a keyboard. He uses the computer to write letters, to maintain records of his work and clients, and to do complex calculations. It also opens doors for him, and operates household appliances.

Another device uses a computer to scan printed text, and translate it into voice output. In this way, blind people can read any book or magazine, rather than the very small number which have been produced in Braille. Voice output promises to make the facilities of the new technology available to the blind.

These are just two examples of how the new technology is already helping the disabled to participate more fully and effectively in society. Here is an undoubted benefit of the microelectronic revolution.

(From *Society and the New Technology*, by Kenneth Ruthren.)

Fig.14.14 This disabled man is communicating with his computer by sucking and blowing through a tube. The computer controls all things in this man's home: the lights, heating, TV, radio, telephone, etc. So he is able to turn on his radio, for example, simply by sucking or blowing through a tube.

Chapter 15 **Yet more questions**

1. The diagram (Fig.15.1) shows a bistable circuit in one of its stable states.

a What effect does pressing switch P have? Explain your answer.

b What effect does pressing switch Q have? Explain.

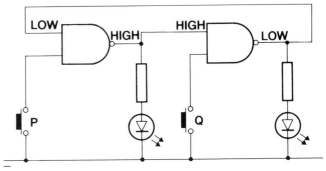

Fig.15.1

2. A student sets up the circuit in Fig.15.2, with the switch S open.

a What will the voltage levels be at each of the NAND circuit outputs A, B and C?

b What do you think will happen when S is closed if each of the NAND circuits takes a little time to change?

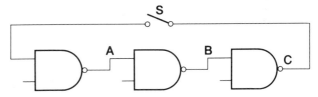

Fig.15.2

(You might like to set this up and try it, perhaps using more NAND circuits. Comment on the number of NAND circuits you might use in the chain to get the effect.)

3. Using NAND gates as the 'building blocks', draw the arrangement you would use for

a an AND circuit,

b an OR circuit,

c a bistable circuit.

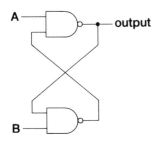

A
output
B

Fig.15.3

4. The circuit in Fig.15.3 appeared in an electronics book.
a Work out what circuit it is.
b Suppose the output of the circuit is high. What would you do to make it low? Explain your answer.

5. In Fig.15.4, what happens to the voltage levels at P and Q as the SPDT switch is moved from position 1 to position 2? How are the NAND circuits behaving? How does this circuit's behaviour differ from that of a bistable?

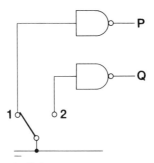

Fig.15.4

6. A student sets up the circuit shown in Fig.15.5(a). First she makes input A low, then input B low and finds that the LED is lit. Next, she makes B high, keeping A low, and finds that the LED is still lit. Then she makes A high, keeping B high, and again the LED is lit. Finally, she makes B low, keeping A high, and the LED is not lit.

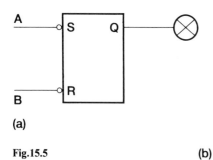

A
S Q
B
R
(a)

Fig.15.5

(b)

B	A	Q
0	0	1
0	1	0
1	0	1
1	1	1

A
Q
B
(c)

The student makes up the truth table of Fig.15.5(b) and concludes that the RS bistable is no different from the circuit of Fig.15.5(c). What would you do to show her that the conclusion is incorrect? What would you say to the student about the way the bistable behaves?

110

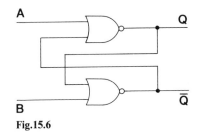

A

B

Q

Q̄

Fig.15.6

7. Use the truth table for a NOR gate to show that the circuit of Fig.15.6 is a bistable circuit if the inputs A and B are at the low voltage level. What would you do

a to set the bistable (that is, make Q=1),

b to reset the bistable (that is, make Q=0)?

8. In the circuit of Fig.15.7, the LED indicator is lit.

a What are the voltage levels at the outputs of AND gates 1 and 2?

b Switch R is pressed and released. What changes, if any, occur?

c Switch S is then pressed and released. What changes, if any, now occur?

d Is this circuit a bistable?

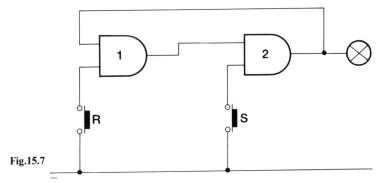

Fig.15.7

9. A student sets up the circuit of Fig.15.8 and claims that it behaves as an automatic light which is on at night and off during the day. Is the claim correct? If not, what change would you make for the circuit to work as intended? Explain how the circuit should work.

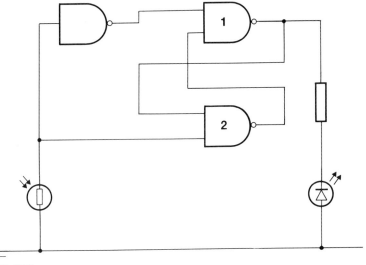

Fig.15.8

10. A NAND gate is closed if any one of its inputs is low. Explain how the bistable circuit of Fig.15.9 works in terms of gates.

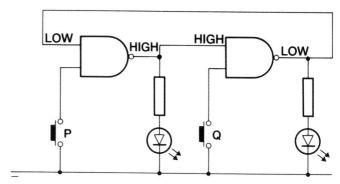

Fig.15.9

11. In an effort to make a latched burglar alarm, a student connects up the circuit of Fig.15.10. Explain why this circuit will not work as an alarm.

What *one* change in the circuit will produce a successful burglar alarm?

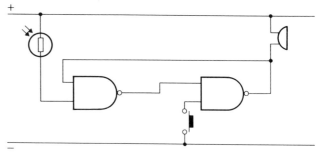

Fig.15.10

12. In the circuit of Fig.15.11, the switches are operated so that the voltage levels at R and at S change with time as shown in the graphs. Draw graphs to show how the voltage level at Q changes with time if it starts at (i) high, (ii) low.

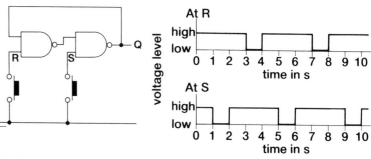

Fig.15.11

13. Bistable circuits can be made from NAND gates or from NOR gates, but not from AND gates or from OR gates. Explain why this is.

14. The circuit shown in Fig.15.12 is part of one used in a quiz-game to identify the first contestant to press an answer switch. The first to press causes his LED to light and prevents the LED of the other contestant from lighting. Initially, both contestants' LEDs are off, that is, H is low and so is H', the corresponding point in the other contestant's circuit.

a What are the voltage levels at the points J and J'?

b What are the levels at the points A, B, C, D, E and F *before* either contestant presses an answer switch?

c What are the levels at the points A, B, D, E, F and H when the answer switch of Fig.15.12 is pressed?

d Explain why the other contestant cannot now light the LED in his circuit.

e What is the purpose of the connection between G and G''?

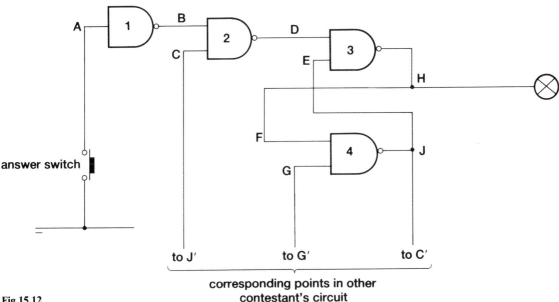

Fig.15.12

corresponding points in other
contestant's circuit

Chapter 16 **Coding**

Communication is all about sending messages. Drum beats, smoke signals, coloured flags at the masthead of ships and flashing lights have all played their part in the sending of messages. In each case it was necessary for there to be a code which was understood both by the sender and the receiver, and it was also necessary for the sender to put the message into code and for the receiver to decode it. In this chapter we will be thinking about coding and in particular how the binary code is used in electronics work.

The binary code

No doubt you are already familiar with counting in the binary system. In this the ordinary decimal numbers 1–10 become binary numbers as follows:

Decimal	Binary
0	0000
1	0001
2	0010
3	0011
4	0100
5	0101
6	0110
7	0111
8	1000
9	1001
10	1010

and so on for higher numbers.

In our electronics work so far, we have been concerned with two states; an output can be either high or low and we have come to represent those states by 1 or 0. All our work with logic gates has been concerned with switching from one state to another. It is because there are just these two states that it is much easier in electronics work to convey messages and to store information using a binary pattern.

Thus, a high level can represent the binary digit 1 and a low level the binary digit 0. The term **binary digit** is usually shortened to **bit**. A number such as 1101 is said to be a 4-bit binary number.

Experiment 16.1 Sending messages using a 4-bit binary code

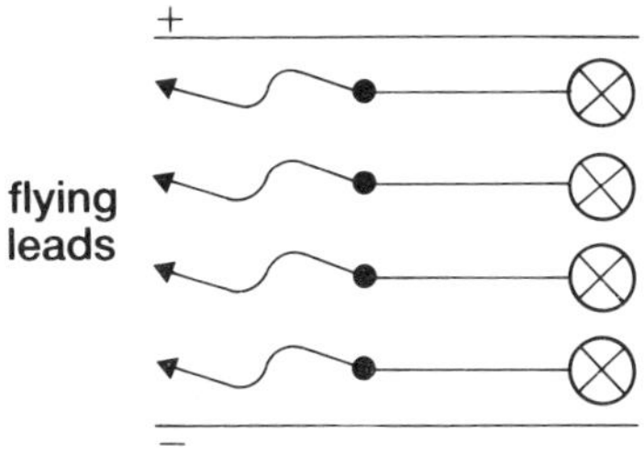

flying leads

Fig.16.1

Connect a power supply to the LED indicator module and a flying lead to each of the four inputs (Fig.16.1).

The LEDs are turned on or off by taking the flying leads at their inputs either high or low.

When an input is high and the corresponding LED is on, it represents the binary digit 1. When an input is low and the corresponding LED is off, it represents the binary digit 0.

1. Together with your partner, invent a 4-bit binary code for sending messages. Write your code in a table like that shown below. A few words have been added to start you off, but you can replace them by your own if you prefer.

Code	Word
0000	
0001	what
0010	the
0011	day
0100	is
0101	it
0110	Monday
....	......

2. How many words can you represent with a 4-bit pattern? Why is 0000 not used to represent anything in your code?

3. Give a copy of your code to another group, and then send them a message using the LED indicators. Send the message, word by word, by lighting the agreed 4-bit binary pattern for each word.

4. Apart from a copy of your code, what other information did you have to give the other group before they were able to decode your message?

Although very simple, this is an important experiment. It shows that a 4-bit binary pattern can be used to represent non-numerical information. It also shows that it is necessary to have a code if information is to be represented by a binary pattern.

With a 4-bit code it is possible to represent sixteen (2^4) words (including 0000 to represent a blank), since there are sixteen distinct binary patterns. How many words can be represented with a 5-bit code? With a 6-bit code?

An 8-bit microprocessor is so called because it recognises an 8-bit binary code. This means it can recognise up to $2^8 = 256$ different instructions.

115

⊗ ON

⊗ ON

⊗ OFF

⊗ ON

Fig.16.2

The most significant bit and the least significant bit

Suppose that, in Experiment 16.1, you wished to send a scale-of-ten number by a binary code. Suppose the LEDs were arranged vertically and were as in Fig.16.2.

Does that pattern represent the binary number 1101 (reading down) or does it represent 1011 (reading up)? It makes a difference, for 1101 in binary stands for 13 in scale-of-ten, and 1011 stands for 11 in scale-of-ten. You probably came across this problem in using your code to send messages in Experiment 16.1. When using a binary pattern to represent numbers (or words), it is vital to know which way round to read the pattern.

In the scale-of-ten number 5327, the 5 stands for 5 thousands or 5×10^3, the 3 for 3 hundreds (3×10^2), the 2 for 2 tens (2×10^1) and the 7 for 7 units (7×10^0). 5 is referred to as the most significant digit because it is the number of thousands and so is the most important. 7 is the least significant digit because it is only the number of units.

In a *binary* number, the digits stand for the number of powers of 2, and the one which represents the highest power of 2 is called the most significant bit (MSB). The one representing the smallest power of 2 ($\times 2^0$) is the least significant bit (LSB). In this book, the LSB will always be shown on the right when binary patterns are written horizontally. This is the normal convention for numbers. Thus, for 1101:

MSB			LSB
2^3	2^2	2^1	2^0
1	1	0	1

represents the number 13 in the scale-of-ten (decimal) notation.

Encoding and decoding

The words encoding and decoding are important ones. *Encoding* means that information is put into a binary pattern. In the case of numbers, it usually means changing from a scale-of-ten (decimal) to a binary notation.

Decoding is the reverse process. The information is extracted from the binary pattern. In the case of numbers, this usually means changing from binary to scale-of-ten (decimal).

Seven-segment displays

Seven-segment displays have become very common these days. For example, you see them on watches and on pumps at petrol stations (Fig.16.3). By means of seven lines, it is possible to represent all the digits from 0 to 9.

(a) Digital watch

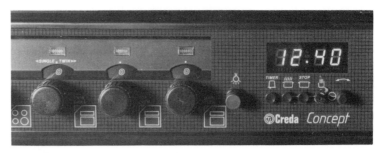

(b) Cooker

Fig.16.3 Uses of seven-segment displays.

(c) Video recorder

In your apparatus you are provided with a seven-segment display module (Fig.16.4). The module also includes a decoder 'chip', but you will not need to use this in the first experiment.

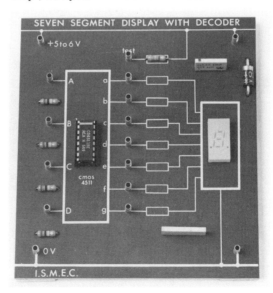

Fig.16.4 Seven-segment display module with decoder.

117

The display will be one of two types, depending on the equipment you are using. One type is called a 'common anode' display (Fig.16.5), while the other type is known as a 'common cathode' display (Fig.16.6). Each has seven segments labelled a to g in the diagrams.

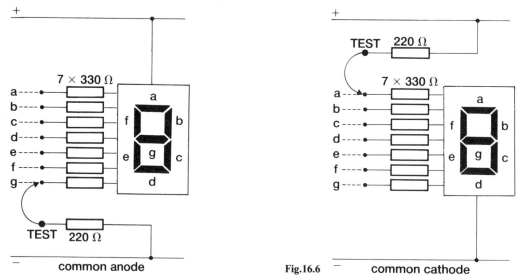

Fig.16.5 — common anode   Fig.16.6 — common cathode

A single segment behaves like an ordinary LED, and lights up when current flows through it. The side at which the current enters the LED is called the anode and the other side is called the cathode (see Fig.16.7).

anode   cathode

Fig.16.7

The difference between the two versions is that all the anodes inside the *common anode* seven-segment display are joined to a single lead, which is connected to the *positive* supply rail. Inside the *common cathode* display, all the cathodes are joined to a single lead, which is connected to the *negative* supply rail.

Experiment 16.2   Investigating a seven-segment display

1. Connect a power supply to the seven-segment display module and a flying lead to the point marked TEST on the display module. Touch the other end on to the pins marked a to g, one at a time. What happens?

2. Copy the table on the left and complete it to show which pins you would touch to display the digits 0 to 9.

Digit	Segments which are lit
0	
1	
2	
3	
4	
5	
6	
7	
8	
9	

The decoder

The integrated circuit on the module is a specially designed chip which takes a binary number on the input side and decodes it into a signal which lights up the appropriate segments on the seven-segment display.

To express the digits 0 to 9 in binary requires a 4-bit binary pattern, so that four inputs are necessary. A is the input for the LSB and D is the input for the MSB (see Fig.16.8).

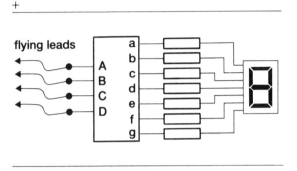

Fig.16.8

So when the binary number 0000 is applied on the input side, the segments a, b, c, d, e, and f light up. When the binary number 0001 is applied to the DCBA inputs, only the segments b and c light up, thereby displaying the digit 1. The other binary inputs are converted in a similar way so that they are shown in scale-of-ten (decimal) on the display.

The integrated circuit used is often called a BCD-to-7-segment decoder/driver, where BCD stands for 'Binary-Coded-Decimal'. The input has to be a decimal digit in the binary code (0000 to 1001), and the decoder decodes the input to drive a seven-segment display.

Experiment 16.3   Using the seven-segment display with its decoder

In this experiment you will again use the seven-segment display module. However this time the four decoder inputs will be used to light the display.

1. Connect a power supply to the display module and flying leads to the inputs A,B,C and D as in Fig.16.8.

2. Use the flying leads to take the decoder inputs high or low and complete the following table by writing in the last column the digit displayed. As usual a high input represents the binary digit 1, and a low input the binary digit 0.

D	C	B	A	
0	0	0	0	
0	0	0	1	
0	0	1	0	
0	0	1	1	
0	1	0	0	
0	1	0	1	
0	1	1	0	
0	1	1	1	

D	C	B	A	
1	0	0	0	
1	0	0	1	
1	0	1	0	
1	0	1	1	
1	1	0	0	
1	1	0	1	
1	1	1	0	
1	1	1	1	

3. When the digit 1 is displayed, segments b and c of the display are lit. Which outputs of the decoder must then be low, and which high?

As there are four inputs to the decoder, it can receive 2^4 or 16 separate messages. But the seven-segment display only needs to respond to ten signals, those corresponding to the binary numbers 0000 to 1001 (in other words the digits 0 to 9). In practice, BCD decoder/drivers may or may not respond to the numbers beyond 1001 (that is, from 1010 to 1111) depending on the particular integrated circuit used. With some, the display will be blank beyond 1001. With others, strange patterns will be lit which do not correspond to any recognisable number. Do not worry about this: the important thing is that the digits 0 to 9 appear correctly!

Symbol for a seven-segment display with decoder/driver

Unfortunately there is no officially recognised symbol for such a system. For convenience in the rest of this book, we will use the simplified drawing shown in Fig.16.9.

Fig.16.9

Experiment 16.4 A 3-line to 8-line decoder

This experiment is only possible if you have a 3-line to 8-line decoder module (see Fig.16.10). Because of all the sockets it is rather an expensive module to make! Furthermore, it needs eight indicators or two LED indicator modules at its outputs. However there are some interesting applications using the module which you may like to try later in the course.

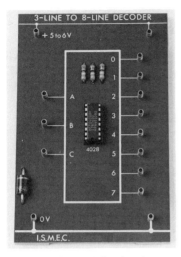

Fig.16.10 3-line to 8-line decoder.

If you have the module, set it up as shown in Fig.16.11 with flying leads and two LED indicator modules.

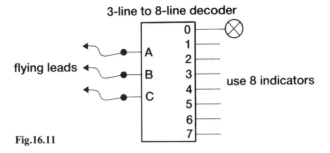

Fig.16.11

1. By taking the inputs high and low, complete the table below.

C	B	A	Indicator lit
0	0	0	
0	0	1	
0	1	0	
0	1	1	
1	0	0	
1	0	1	
1	1	0	
1	1	1	

2. Write a sentence summarising the behaviour of this type of decoder.

The input to this decoder is a 3-bit binary pattern (000 to 111). This means that eight different input patterns are possible (2^3). There are eight output lines and each possible input pattern causes a different output line to go high. When this happens the other outputs stay low.

With this type of decoder we say that the 3-bit input pattern is 'completely decoded'. This means that a signal is produced on a separate output line for each possible binary input pattern. The signal of course is one of the outputs going from low to high.

With only two inputs, there are four possible binary patterns (00,01,10,11), so that a complete decoder would require four outputs (2^2) and would be called a 2-line to 4-line decoder. For the complete decoding of a 4-bit binary pattern, sixteen (2^4) outputs are required. In general, an N-bit binary input requires 2^N outputs for complete decoding. Such decoders can be built using NAND gates, as the next experiment shows.

Experiment 16.5 Making a 2-line to 4-line decoder from NAND gates

First consider the pattern of gates shown in Fig.16.12. On the left are two inputs labelled X and Y which are inverted by NAND gates 1 and 2 to produce $\overline{X}$ and $\overline{Y}$. On the right is an AND gate, formed from NAND gates 3 and 4, with two inputs A and B.

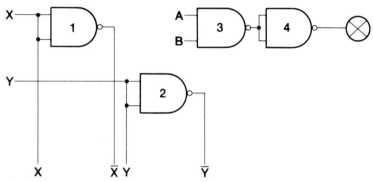

Fig.16.12

1. If $X = 0$ and $Y = 0$, how would you connect A and B in order to light the LED? How would you connect A and B to light the LED if $X = 0$ and $Y = 1$? How would you do it if $X = 1$ and $Y = 0$? And if $X = 1$ and $Y = 1$? Remember that, each time, A and B must both be high to cause the LED to light. Now build the circuits and check your predictions.

2. Use your results from above to draw a circuit diagram of a 2-line to 4-line decoder. You will, of course, need four AND gates. If the NAND gates are available, build the decoder.

3. NAND gates with three inputs are also available in integrated circuit form (the output is low only when all three inputs are high). Try to draw a circuit diagram showing how a 3-line to 8-line decoder could be built from 2-input and 3-input NAND gates.

4. A 3-line to 8-line decoder can be built using 2-input NAND gates only. How many of these gates would be required? Explain your reasoning.

A 3-input decoder would require a NAND gate used as an inverter for each of the inputs. Each output would require a 3-input AND gate, that is, a 3-input NAND gate with another NAND gate to invert the output. Thus $3 + (8 \times 2) = 19$ NAND gates are necessary for a 3-line to 8-line decoder, eight 3-input NANDS and eleven inverters.

If only 2-input NAND gates are available, the 3-input NAND gates have to be replaced by the circuit of Fig.16.13. Now the 3-line to 8-line decoder needs 35 NAND gates – and they can easily be accommodated on an IC chip!

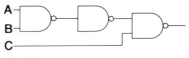

Fig.16.13

Project   A one octave music box with a three-key keyboard

In the circuit shown in Fig.16.14, the eight outputs of a 3-line to 8-line decoder are connected to the eight inputs of a music box module (Fig.16.15). The normally high decoder inputs can be taken low using three push-button switches. These switches make a three-key keyboard.

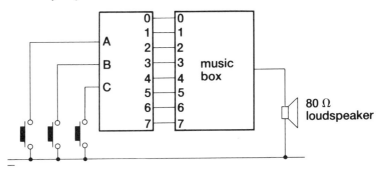

Fig.16.14

1. Explain why just three keys can be used to generate a full octave (eight notes).

2. Try playing scales and tunes!

Fig.16.15

Background reading

The ASCII code

A *byte* is simply another name for an 8-bit binary pattern. The memories found in many microcomputers are such that a single byte of information can be stored in each memory location. But what do we mean by a 'byte of information'?

Before the pattern which is stored can take on some meaning for a human being and become information, an agreed code is necessary. You learnt about inventing codes for binary patterns in Experiment 16.1. You probably found in that experiment that different groups of people invented different codes, and that you could only understand another group's messages if you knew their code. Obviously if all the groups in the class could have agreed on the same code, then everyone would have been able to understand everyone else's messages. Such a code would have been a *standard code* for the class.

Something like this happens in the world of computers. There is a standard code known as the American Standard Code for the Interchange of Information (ASCII for short) which manufacturers throughout the world use in their machines. This enables the machines to communicate with each other (perhaps using telephone lines or communication satellites) and also with devices such as printers which use the same code.

The ASCII code uses byte-long data patterns to represent what are called *alphanumeric* characters, that is, the letters of the alphabet, the numbers 0 to 9 and various punctuation marks. Each character has its own byte-long code.

A simple message in ASCII code is given below.

H	0100	1000
E	0100	0101
L	0100	1100
L	0100	1100
O	0100	1111
T	0101	0100
H	0100	1000
E	0100	0101
R	0101	0010
E	0100	0101
!	0010	0001

Note that with an 8-bit code of this kind, there are 2^8 or 256 distinct binary patterns. This, of course, is more than enough to represent all the alphanumeric characters.

If you want to see the full ASCII code, you should try looking in the manual of a microcomputer. The code is often included in an appendix.

Chapter 17

The pulser, the astable and the clocked bistable

In this chapter two new modules will be introduced, both of which will provide opportunities to do a wider range of useful jobs. The first of these is the pulser/astable module shown in Fig.17.1. This module has two *independent* sections. The first is known as a *pulser*, and has one output socket and one switch. Below this is an *astable* section with two output sockets and two switches. The two circuits for the pulser and astable sections are completely separate.

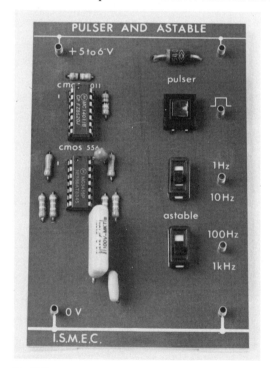

Fig.17.1

Experiment 17.1   Using an oscilloscope to observe the voltage outputs from the pulser/astable module

In this experiment you will use the pulser/astable module as shown in Fig.17.2. Connect the lower astable output to an oscilloscope and move the switch alongside the output socket to the 100 Hz position. Make sure that the module is connected to its power supply.

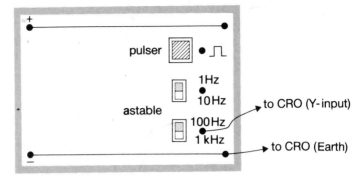

Fig.17.2

1. Adjust the oscilloscope until a stationary trace is obtained. Sketch the waveform you see.

2. What is
a the height of the waveform in volts,
b its period in milliseconds,
c its frequency?

3. Move the lower switch to the 1 kHz position and repeat the above.

4. Transfer the Y-input lead from the lower socket to the 1 Hz/10 Hz socket. Observe the trace with the upper astable switch in each of its two positions. Measure the height of the waveform in each case, and its period in seconds. Calculate the frequencies.

5. Transfer the Y-input lead to the pulser socket (⎍). Press the pulser switch down, hold it down for a moment or two and then release it. What happens?

6. Try connecting LED indicators to the outputs instead of the CRO.

Like other digital circuits, the output of an astable can be high or low. However, in the case of an astable the output continually and automatically switches between these two states. The output is stable in neither state – hence the name *astable*.

Other names are sometimes used to describe such a circuit. These include square wave oscillator, square wave generator, and (free-running) multivibrator. In this book the term astable will be used.

This course is concerned with *how* astables behave, not with *why* they behave in this way. So we shall not be concerned with the circuit inside the astable.

127

Fig.17.3

The astable module produces a 'square' output voltage waveform similar to that shown in Fig.17.3. It will be seen that the time for which the output is high is about the same as the time for which the output is low. There are four different frequencies of the output: 1 Hz, 10 Hz, 100 Hz and 1 kHz. The term *clock signal* is often used to describe an output like this. Clock signals from the astable unit will be used often to provide timing signals for a number of different electronic circuits later in this chapter and in the rest of this book.

The pulser output is different. It goes high when the switch is pressed and returns to low when the switch is released. It provides a convenient way of producing a single pulse and controlling by hand how long it lasts. You may wonder why you cannot use an ordinary switch for this and why there has to be a special circuit. The reasons are given in Appendix A on page 176.

The clocked bistable

The other module to be introduced in this chapter is the clocked bistables (Fig.17.4). Like the bistable built from NAND gates earlier, there are SET and RESET inputs (see Fig.11.10 on page 88), and Q and $\overline{Q}$ outputs. The only difference is that there is now an extra input, namely the clock input. The symbol to be used for the RS clocked bistable is shown in Fig.17.5.

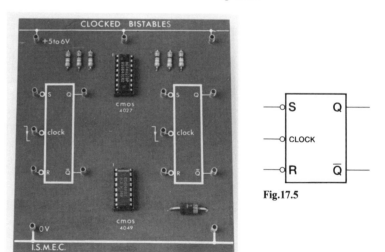

Fig.17.4

Fig.17.5

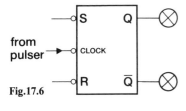

from pulser →

Fig.17.6

Experiment 17.2 Investigating a clocked bistable

The circuit is shown in Fig.17.6 (remember to connect a power supply). Connect each of the outputs, Q and $\overline{Q}$, to an LED indicator. Attach flying leads to the inputs S and R.

1. First, without the pulser connected, check that the SET and RESET inputs behave in the same way as with an ordinary bistable. If the SET input (S) is taken briefly low, the bistable enters the state with Q high (Q = 1) and $\overline{Q}$ low ($\overline{Q}$ = 0) – the SET state in which Q has been *set* to 1. When the RESET input (R) is taken briefly low, the state with Q = 0 and $\overline{Q}$ = 1 is entered – the RESET state, in which Q has been *reset* to 0.

2. Now it is necessary to find out what the clock input does. Remove the flying leads and leave the SET and RESET inputs unconnected. Connect the clock input to the pulser. Apply pulses one at a time to the clock input. What happens?

3. Does the bistable change state when the pulse is applied ('the rising edge of the clock pulse') or when it is switched off ('the falling edge')?

4. Now try clocking the bistable with the 1 Hz output from the astable part of the pulser/astable module. What is the frequency of the pulses at the Q output?

Note that if the clock input is not used, the new bistable behaves in exactly the same way as the NAND gate version.

However, when the pulser is connected to the clock input and the SET and RESET inputs are not used, the bistable changes state every time the pulser is operated. In other words, the bistable changes state every time a complete low–high–low clock pulse is applied to its clock input.

Note that the actual change of state only occurs on the *falling edge* of the clock pulse. In other words, the change occurs when the pulse switch is released and not when it is pressed. The changes in Q and $\overline{Q}$ are summarised in Fig.17.7, called a *timing diagram*.

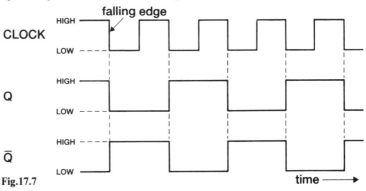

Fig.17.7

129

Notice that one complete low–high–low change of Q is produced for every *two* clock pulses at the input. In other words, the bistable produces one output pulse for every two input pulses.

It is not necessary to know the details of the circuit in the IC which enables this to be done. You can think of it as a sort of steering circuit which directs the clock pulses alternately to the SET and RESET inputs.

Experiment 17.3 Constructing a binary counter from clocked bistables

Connect four clocked bistables together as in Fig.17.8. Connect an LED indicator to each of the Q outputs, labelled A, B, C and D in the diagram. Connect all the RESET inputs together and use a flying lead to turn all the indicators off. In other words, make all the Q outputs low.

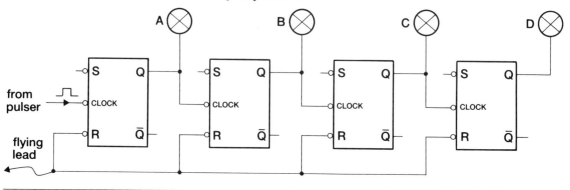

Fig.17.8

1. Apply pulses one at a time to the left-hand bistable using the pulser. Complete the following table, noting that the states of indicators ABCD are entered in the order DCBA in the right-hand columns in the tables below.

Pulse number	D	C	B	A		Pulse number	D	C	B	A
0	0	0	0	0		8				
1						9				
2						10				
3						11				
4						12				
5						13				
6						14				
7						15				

Note that the right-hand columns represent the number of pulses as a 4-bit binary number. In other words, the four clocked bistables act as a 4-bit binary counter with A indicating the LSB.

2. Replace the pulser by a 1 Hz clock signal from the astable. Watch the system continually counting from 0 to 15 in binary.

3. Remove the astable and reset all outputs to zero (all indicators off). Use the pulser to *apply 16 pulses* one at a time to the left-hand flip-flop. Complete the following:

Number of clock pulses at input = 16
Number of pulses at output A　=
Number of pulses at output B　=
Number of pulses at output C　=
Number of pulses at output D　=

4. If a clock signal of frequency 1 Hz were applied to the left-hand clock input, what would you expect the frequency of the pulses produced at the D output to be? Use the 1 Hz output from the astable and a stopwatch to check your prediction.

When clocked bistables are linked to form binary counters, it is usual to feed the external clock pulses into the left-hand bistable. The Q output of this bistable is therefore the least significant bit (LSB) and the Q output of the right-hand bistable the most significant bit (MSB). This, of course, results in a display in the reverse order to which binary numbers are usually written.

This experiment shows how clocked bistables can be linked together to count pulses in binary. With four bistables, counts from 0 to 15 (binary 0000 to 1111) can be displayed. To extend the maximum count, more bistables are added. The maximum count with N bistables is $2^N - 1$.

The timing diagram (Fig.17.9) shows how each of the Q outputs changes as clock pulses are fed into the left-hand bistable.

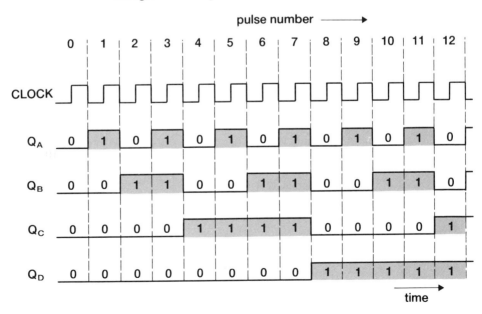

Fig.17.9

131

Suppose each of the Q outputs is initially low. Each bistable changes state on the falling edge of the pulse at its own clock input. Q_A changes on the falling edge of the pulse at its clock input; Q_B changes on the falling edge of Q_A; Q_C on the falling edge of Q_B; Q_D on the falling edge of Q_C.

Divide-by-*N* counters

You were asked in part 3 of Experiment 17.3 to look at the counting in a slightly different way. For every sixteen pulses at the input, Q_A pulsed eight times, Q_B pulsed four times, Q_C pulsed twice and Q_D once. Each bistable divides the clock frequency by 2, so that, with four bistables in sequence, the output frequency of the last bistable is one-sixteenth of the input to the first. This is often called a divide-by-16 counter. In general, N clocked bistables in sequence will form a divide-by-2^N circuit.

These counters are extensively used in electronics when it is necessary to have signals which are exact multiples of one another. For example, in one make of computer, an astable of frequency 16 MHz is used as the input to a divide-by-16 counter and the 1 MHz output is used to drive the microprocessor.

Digital watches use an oscillator which generates pulses at a frequency of 32768 Hz. How many bistables must be in sequence for the last one to generate a pulse every second? The answer is 15.

Experiment 17.4 Constructing a binary down-counter

In Experiment 17.3, the counter counted up from 0 to 15 in binary: it was a binary 'up-counter'. In this experiment, the counter will count downwards from 15 to 0: so it is a binary 'down-counter'.

Change the circuit used in experiment 17.3 so that the LEDs are driven by the $\overline{Q}$ outputs. Leave the Q outputs connected to the clock inputs. Use the flying lead connected to all the RESET inputs to turn all indicators on (all the $\overline{Q}$ outputs high).

1. Apply pulses one at a time to the left-hand bistable.

Pulse number	D	C	B	A		Pulse number	D	C	B	A
0	1	1	1	1		8				
1						9				
2						10				
3						11				
4						12				
5						13				
6						14				
7						15				

2. Complete the tables on page 132. Note that the states of the indicators are entered in the order DCBA in the right-hand column.

Note that at each pulse, 1 is subtracted from the count shown by the LED indicators. It is indeed a binary down-counter.

Applications using the RS clocked bistable

Project 17(a) Automatic hoist
The hoist is a motor-driven machine which lifts sacks from the ground floor of a warehouse to an upper floor. The machine is started by pressing a switch on the ground floor and stops automatically when the load interrupts a light beam at the upper level. Design a circuit to do this. Suggest a way of returning the hoist to ground level.

Project 17(b) Lamp flasher
Use a clocked bistable, the NAND module, two LEDs on the indicator module and the 1 Hz pulses from the astable to produce a circuit which will light two LEDs, X and Y, in the sequence:

only X on → both off → only Y on → both off → and so on.
(*Hint*: draw a timing diagram showing the clocking pulses and the voltage levels at Q and $\overline{Q}$.)

Project 17(c) Automatic light-buoy
A warning buoy in a shipping lane is to have a light which flashes. It is to be on for one second and then off for three seconds. Design a circuit which will do this, using the RS clocked bistable module, the NAND gate module, an LED on the indicator module and the 1 Hz output from the astable. Then change the circuit so that the LED is on for 3 seconds and off for 1 second.

Project 17(d) A reversible counter
In Fig.17.10 the box X is an arrangement of NAND gates such that the LED indicates the voltage level at Q if P is low and the

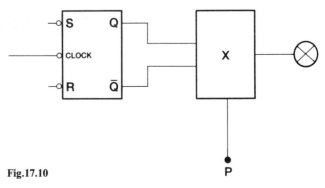

Fig.17.10

133

voltage level at $\overline{Q}$ if P is high. Devise a circuit to do this. (*Hint*: the LED lights for Q = 1 or for $\overline{Q}$ = 1, but P = 0 closes the route from $\overline{Q}$ and opens that from Q.)

If the circuit was used between the bistables in a counter, then with P = 0 the circuit would be an up-counter and with P = 1, a down-counter. Of course, the next bistable would be clocked by the Q output.

Project 17(e) Automatic gating circuit
A circuit is to be used as a gate which will be opened by the first pulse to arrive at the input, so allowing all the following pulses through until the circuit is reset. Use an RS clocked bistable and a NAND gate to do this. The pulser is a suitable source of pulses with an LED as indicator.

Background reading

Gates
An easy way of understanding why logic gates are called 'gates' is to think of one of the two inputs as a *control*. For an AND gate, the truth table can then be rewritten in the way shown below.

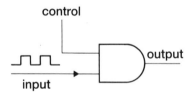

Fig.17.11

Control	Input	Output
low low	low high	low low
high high	low high	low high

Imagine a series of high and low pulses arriving at the other input as shown. As long as the control input is low, the output is always low whatever the state of the input. The gate is *closed*.

But if the control is high, the output is the same as the

other input. In other words, the gate is *open*. It is behaving just like a real gate, sometimes open and sometimes closed. When closed, nothing 'passes through'; when open, the output is the same as the input, the pulses 'pass through'.

NAND, OR and NOR gates can be looked at in the same way. A NAND gate is open when its control input is high as with an AND gate, but the pulses arriving at the other input are inverted when they arrive at the output (high voltages become low and low voltages high).

OR and NOR gates are both open when their control inputs are low, but the output pulses are inverted in the case of the NOR gate.

The simple set-up below can be used to show the gating action. You might like to try it with your electronics modules. The first NAND gate, followed by the second as an inverter, form an AND gate.

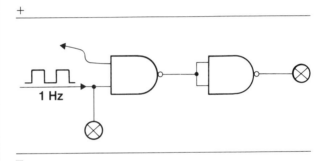

Fig.17.12

When the control input is high, pulses will pass through the AND gate and the two LEDs will flash on and off together. The gate is *open*. If the control input is low, the LED at the output of the AND gate will be permanently low. The gate is *closed*.

If a single NAND gate is used (without the inverter) and the control is high, the output LED will flash on and off, but it will be on when the other is off, and vice versa. When the control is low, the output LED will be permanently on. Other gates can be investigated in the same way.

Chapter 18 **Counting circuits**

In Experiment 17.3 in the previous chapter, clocked bistables were joined together to make a binary counter. Four bistables were needed to make a 4-bit binary counter. Instead of using a chain of bistables every time we wish to count pulses, it is far more convenient to use an integrated circuit to do the job. The binary counter module, shown in Fig.18.1, uses a single 4-bit binary counter IC. This contains the necessary four bistables joined in sequence. The module has a clock input, and four outputs corresponding to the A, B, C and D outputs in Fig.17.8. There is also a reset input. The first experiment puts this new module to use.

Fig.18.1

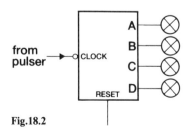

Fig.18.2

Experiment 18.1   Counting pulses

Connect the pulser to the input of the 4-bit binary counter module. Link each of the four output lines to one of the inputs on the LED indicator module, as in Fig.18.2.

Connect a flying lead to the RESET input. If any of the indicators are lit, use the flying lead to take the RESET input briefly *high*. Note that the counter can be reset to zero at any time by taking this RESET input *high*.

1. Apply pulses to the counter one at a time. Does the display change on the rising or falling edge of the clock pulse?

2. Send 16 pulses to the counter and check that the indicators correctly display the number of pulses in binary.

3. Now replace the pulser by the 1 Hz output signal from an astable unit. Watch the system repeatedly cycle from 0000 through to 1111 and back to 0000 again.

4. With the 1 Hz signal still connected, remove indicators A, B and C. What is the frequency of the output signal at D? Is this what you expect?

Experiment 18.2   Linking the counter to a decoder and seven-segment display

The experiment above made it possible to count pulses in binary. By combining the counter with the decoder/seven-segment display module it is possible to count in the scale-of-ten or decimal notation.

Instead of the LED indicators used in the previous experiment, connect the decoder/seven-segment display module as in Fig.18.3.

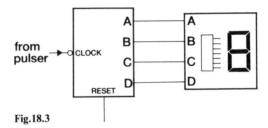

Fig.18.3

1. First reset the counter by taking the RESET input briefly high.

2. Use the pulser to apply pulses one at a time to the clock input of the counter. Copy down the display after each pulse (Fig.18.4).

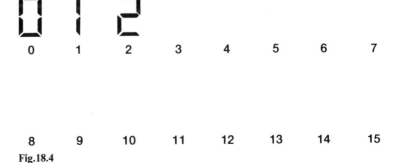

Fig.18.4

137

The binary counter module is a 4-bit counter and it will therefore count from 0000 to 1111, in other words from 0 to 15. As explained earlier, what appears on the seven-segment display will depend on the particular type of module which your school has. The binary numbers 1010 to 1111 will still be counted by the 4-bit binary counter, but the display will either be blank or show patterns which are not decimal digits.

Counting in the scale-of-ten (decimal)

The unsatisfactory thing in the previous experiment was that the display showed the digits 0 to 9 and did not display them again until sixteen pulses had been given to the counter. Clearly the circuit needs resetting immediately after the tenth pulse so that instead of going on to 1010, it resets to 0000.

It has already been seen that the way to reset the binary counter is to make the RESET input momentarily high. How can this be done?

The answer is simple – by using the logic gates with which you are already familiar. The binary number corresponding to ten is 1010. When this number is reached, the outputs of the 4-bit binary counter will have D and B high, and A and C low.

If an AND gate were connected between D and B, the output of the AND gate would be high when D and B were high. If the AND output were connected to the RESET input, that input would then go high and the binary counter would be reset (Fig.18.5).

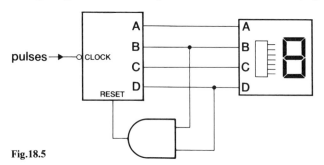

Fig.18.5

Experiment 18.3   Counting in the scale-of-ten

1. Set up the circuit as in Fig.18.6. The two NAND gates together provide the necessary AND gate.

2. Apply a 1 Hz signal from the astable unit to the clock input of the binary counter. Check that the system now counts from 0 to 9, then resets to 0 and starts counting upwards from zero again.

The circuit is now a counter for which there are only ten possible output patterns in the binary code (0000 to 1001). It is called a **b**inary **c**oded **d**ecimal or BCD counter.

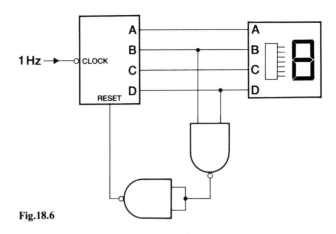

Fig.18.6

There is little need to stress the importance of counting circuits. For example, a petrol pump counts up the number of volume units (one-hundredths of a litre) – and the cost as well – and displays it as a scale-of-ten number. To do that, the counter for the least significant digit (the units) has to clock the 'tens' counter every time it resets from 9 to 0, and the 'tens' counter has to clock the 'hundreds' counter when it resets, and so on.

Experiment 18.4 Linking together two counters to count to 99

The problem in this experiment is to produce a counter which will display counts from 0 to 99 on two seven-segment displays, using two 4-bit binary counters and two seven-segment display modules with decoder/drivers. To find a solution, you will need to use a quad NAND module as well. The astable module will also be needed to supply the pulses to be counted.

You might like to try this experiment as a project. If you do, then here is a clue. It will be necessary for a pulse from the first counter to clock the second counter when the number ten is reached and the first counter resets; and the 4-bit binary counter clocks when its clock input goes from high to low.

Resetting counters at other stages

Counters which reset at a stage other than ten are also needed. In the next section, a digital clock will be studied in some detail and you will find it necessary to have a counter which counts from 0 to 4 and then resets (a scale-of-five counter), and counters in the scale-of-six (reset after reaching a count of 5), and a scale-of-seven counter. In general, a scale-of-N counter counts from 0 to $N-1$ and resets to 0 after the Nth pulse. It is often called a divide-by-N or $\div N$ counter.

Experiment 18.5   Building scale-of-*N* counters

1. Set up the circuit in Fig.18.7 to produce a scale-of-five counter, in other words a counter which counts from 0 to 4 and then resets to 0.

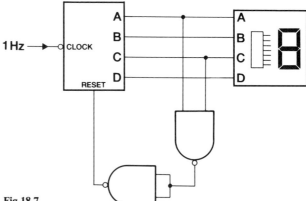

Fig.18.7

2. Change the connections to make a scale-of-six counter.

3. (more difficult) Change the circuit to make a scale-of-seven counter. (*Hint*: you will need to make a 3-input AND gate.)

The digital clock

You can see the growth in the use of electronic systems all around you. There are digital cash registers, calculators and computers, clocks and watches, petrol pumps, digital balances, hi-fi systems, telephone dialling, bank cash dispensers, washing machines, and so on. All use systems built from the building blocks you have been studying, and you may have appreciated that it is not too difficult to design a system once you know what the inputs are, what output is needed and what building blocks are available.

The digital clock is a good example of the use of much of the electronics you have seen up to now. Suppose the clock is to be like one of those used in a video-recorder to programme a recording. The clock output is a display showing the day of the week, whether it is a.m. or p.m., the hour and the minutes, and, perhaps, even the seconds (Fig.18.8). How is that system put together?

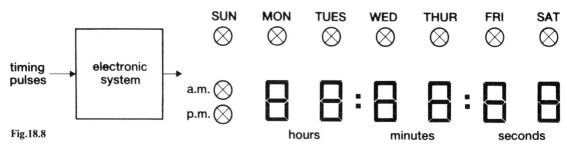

Fig.18.8

140

Suppose the timing pulses come from the 50 Hz mains supply. The displays are driven by a BCD input and decoder/driver, so that the system is essentially a counter, and one which will need to supply pulses at various rates. The seconds counter requires one pulse per second and must reset on reaching 60. Resetting the seconds counter must supply a pulse to the minutes counter, and that in turn must pulse the hours counter every hour, and so on.

Consider the seconds counter. How do you obtain 1 pulse per second from an input pulsing at 50 Hz? You may remember that a scale-of-N counter is also a divide-by-N counter, for the pulse that resets the counter occurs once in N counts. So, 1 pulse per second will be obtained from a $\div 50$ counter fed with 50 Hz pulses. To divide by 50, we must first divide by 10 and then by 5 (Fig.18.9).

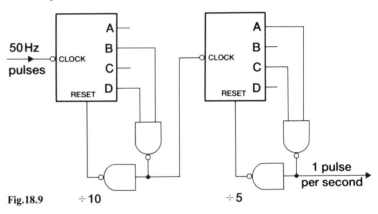

Fig.18.9 $\div 10$ $\div 5$

Now it seems easy; the 1 pulse per second must be the input to a $\div 60$ counter to get 1 pulse per minute. The output from that must be the input to another $\div 60$ counter to get 1 pulse per hour. Then there must be a $\div 12$ counter to show the hours (Fig.18.10).

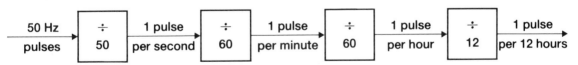

Fig.18.10

Fig.18.11 (over the page) shows a $\div 60$ counter with seven-segment displays to show the units and tens of the pulses being counted. One of the circuits of Fig.18.11 is needed for the seconds and one for the minutes.

But what about the hours counter? A 'units' counter and a 'tens' counter are needed, but since the 'tens' counter is only ever 0 or 1, an RS clocked bistable can be used for that. However, resetting this counter is more of a problem than in the previous cases. The 'units' counter has to be reset not only when it reaches 10, but also when the *total* count has reached 12, so that the time shown 1 minute after 11:59 is 00:00. This requires an OR gate: 'reset' on the

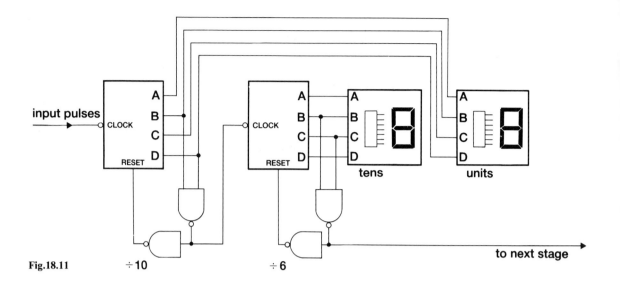

input pulses

to next stage

Fig.18.11 ÷10 ÷6

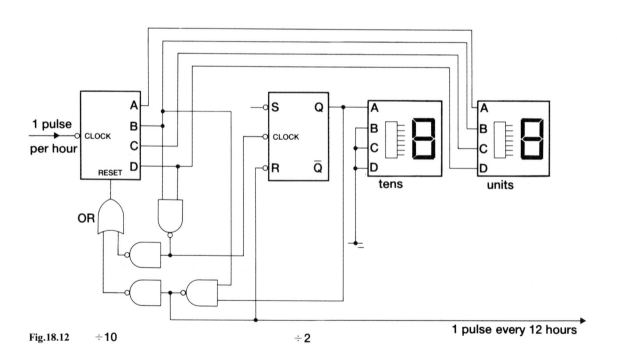

1 pulse per hour

OR

1 pulse every 12 hours

Fig.18.12 ÷10 ÷2

142

'units' counter must be activated when the 'units' count reaches 10 OR when the total count reaches 12. Fig.18.12 shows the circuit needed.

Some simplification in the resetting circuit is possible. If this part of the circuit is drawn using NAND gates, it becomes that shown in Fig.18.13. The gates numbered 1, 2, 3 and 4 are all inverters and gate 1 merely cancels the effect of gate 4, and similarly with gates 2 and 3. These four gates can therefore be omitted to leave the resetting circuit as shown in Fig.18.14.

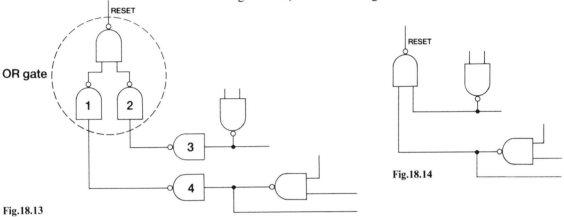

Fig.18.13

Fig.18.14

Already you may have noticed a snag – or at any rate a possible improvement. We are more accustomed to seeing the time change from 12:59 to 01:00 rather than from 11:59 to 00:00. Now we need an hours counter in which the units display becomes 0 when 10 is reached and 1 when 13 is the total. The circuit is that of Fig.18.15.

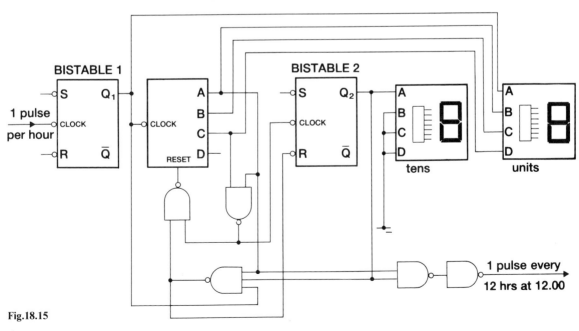

Fig.18.15

143

See if you can puzzle out how it works. The table below gives you hints. It shows the voltage levels at hourly intervals.

Q_1	A	B	C	Q_2	Display	
1	0	0	0	0	0	1
0	1	0	0	0	0	2
1	1	0	0	0	0	3
0	0	1	0	0	0	4
1	0	1	0	0	0	5
0	1	1	0	0	0	6
1	1	1	0	0	0	7
0	0	0	1	0	0	8
1	0	0	1	0	0	9
(0	1	0	1	0)	binary counter resets and clocks bistable 2	
0	0	0	0	1	1	0
1	0	0	0	1	1	1
0	1	0	0	1	1	2
(1	1	0	0	1)	binary counter and bistable 2 are reset (but not bistable 1)	
1	0	0	0	0	0	1

Now the a.m./p.m. indicators and the 'days' counter can receive a pulse every 12 hours. You should be able to understand from Fig.18.16 how the a.m./p.m. indicators are operated and how one pulse in 24 hours is obtained to clock the ÷ 7 counter. To light one of the seven different lamps requires a 3-line to 8-line decoder.

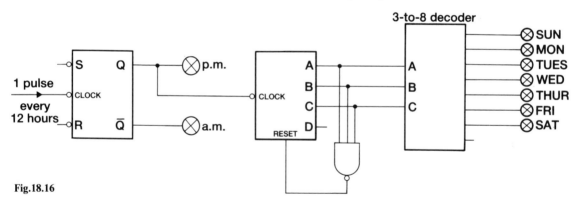

Fig.18.16

What is so remarkable is that all this – and more besides – can be put into one chip! And there is another feature; manufacturers like their products to sell world-wide and some parts of the world use a frequency of 60 Hz for their mains supply rather than 50 Hz. To allow for this, there is one pin of the IC which, if at the high level, causes the clock to register time correctly with a 50 Hz input, whereas if connected to the low level, it registers correctly with a 60 Hz input. See if you can invent a circuit to do that. The answer is in Appendix B on page 179.

Chapter 19

The latch and the dual decade counter

The hold/follow latch

Fig.19.1

The hold/follow latch (Fig.19.1) is an important electronic building brick, which will prove to be useful in your project work. Such a latch can also be thought of as a simple electronic memory, and we will build upon this idea in the next chapter. But first we must find out what the hold/follow latch does.

Experiment 19.1   Using a 4-bit hold/follow latch to store a binary pattern

Set up the circuit shown in Fig.19.2. Use a 10 Hz signal from an astable module to clock the counter.

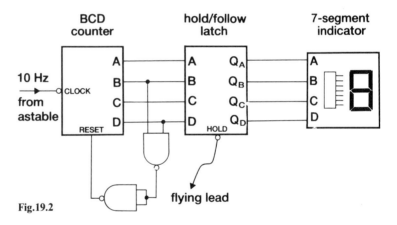

Fig.19.2

1. Check that the seven-segment display is counting the pulses from the astable as it should.

2. The hold input is normally high. Use a flying lead to make it low. What happens to the display?

3. Disconnect the flying lead so that the hold input is once again high. What happens?

4. Write a sentence or two describing the behaviour of the hold/follow latch module.

145

This quick experiment shows the function of the hold/follow latch module. When the hold input is high, the four output lines *follow* the four input lines. If the hold input is taken low, the output lines *hold* the logic levels present on the input lines at the moment the hold input became low and the display no longer changes.

The word 'latch' is used in electronics to mean *store* or *hold*. So, for example, if a binary pattern is 'latched', we mean it is held or stored. A hold/follow latch is often known as a 'storage latch' or 'storage register'.

Another word which is often used is *data*. The 4-bit binary pattern held in the latch is often referred to as stored *data*. In this experiment, the stored data is the binary number present at the counter output when the hold input was taken low.

The hold/follow latch can also be thought of as a simple 4-bit memory. When the hold input goes low, the data present at the latch inputs is memorised (stored) until the hold input goes high again.

The next experiment shows how a very simple latch can be made.

Experiment 19.2   Constructing a 1-bit hold/follow latch from NAND gates and a bistable

Set up the circuit in Fig.19.3 and connect a 1 Hz signal from the astable module to the data input and a flying lead to the hold input.

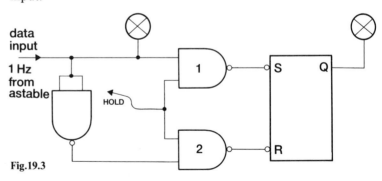

Fig.19.3

1. Take the hold input high. What happens?

2. Now take the hold input low. What happens?

3. Try to explain how the circuit works.

The way the circuit works can be explained as follows. Suppose first, the hold input is high. Then when the data input is high, we have the situation shown in Fig.19.4.

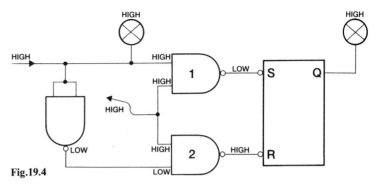

Fig.19.4

When the data input is low, we have the situation in Fig.19.5.

Thus the final output *follows* the input (the two indicators are either both high or both low). When the data input is high, Q is high. When the data input is low, Q is low.

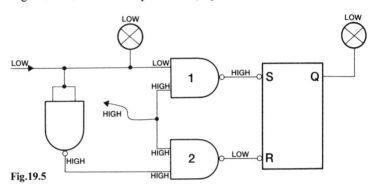

Fig.19.5

(Alternatively, we can look at it from the point of view of 'gates'. Making the hold input high *opens* NAND gates 1 and 2 so that changes in voltage levels at the inputs to 1 and 2 cause changes in the voltage levels at the outputs.)

With the hold input low, we have a different situation. With a high data input, the situation is now as shown in Fig.19.6.

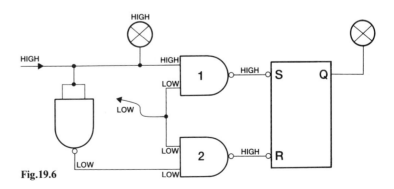

Fig.19.6

147

With a low data input, it is as in Fig.19.7.

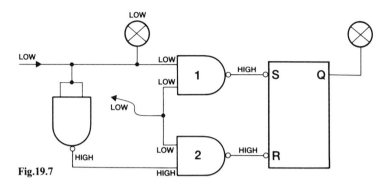

Fig.19.7

In these last two cases, whether the data input is high or low, the outputs of the two NAND gates are both high. But the RS bistable only changes when one of the points S or R goes low, so the output stays as it was when the hold went low. What was then showing is still showing: it is *held*.

(Again we can look at the situation from the point of view of 'gates'. Making the hold input low *closes* gates 1 and 2 and now changes of level at the inputs have no effect on the voltage levels at the outputs.)

The data latch retains the input data for as long as its hold input is kept at the low voltage level, in other words it 'keeps it in its memory'. The hold/follow latch module (Fig.19.1) has four of these 1-bit latches in the IC chip so that four bits of data can be stored or memorised.

The dual decade module

Experiment 18.4 on page 139 was an investigation to see how two single digit counters could be linked together to make a dual decade counter. The solution to the problem is shown in Fig.19.8.

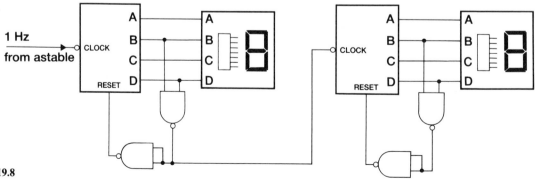

Fig.19.8

148

When the number ten is reached on the first counter, the two NAND gates connected to the D and B outputs ensure that the counter is reset. The output from the first NAND will move from high to low when both D and B go high. By connecting the output of that NAND to the clock input of the second counter, the fall from high to low will be the necessary pulse to clock the second counter. Thus every time the first counter moves from 9 to 0, the second will get a further pulse to count – in other words it will count the tens.

It is convenient to put such a dual decade counter into a single module (Fig.19.9). This saves bench space and connecting leads, and makes circuits needing two counters much easier to set up. The dual decade counter module shown in Fig.19.9 includes hold/follow latches. Fig.19.10 shows the building blocks of the module and the way they are connected together. Note that the counters in this diagram are BCD counters; this means that they count from 0 to 9 and are then reset to zero. Each BCD counter is just a 4-bit binary counter with two NAND gates to reset it to zero, as shown earlier in Fig.19.8. Also, the second counter is shown clocked from the D output of the first counter. This works because the D output goes from high to low when the first counter is reset, which is what we need to clock the second counter. The dual decade counter module does a job which would otherwise

Fig.19.9

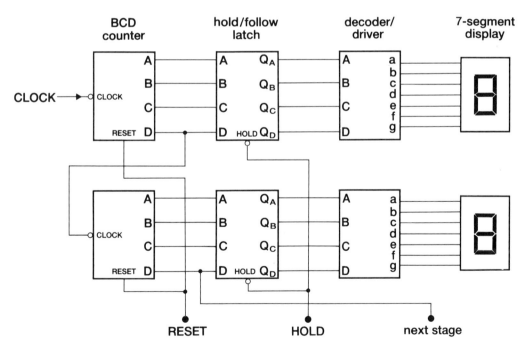

Fig.19.10

require linking together seven separate modules (two 4-bit binary counters, a quad NAND module, two hold/follow latches and two decoder/displays)!

The symbol used in the rest of this book for our dual decade counter module is shown in Fig.19.11.

Fig.19.11

Experiment 19.3   Using the dual decade counter module

Set up the circuit shown in Fig.19.12.

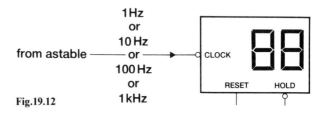

Fig.19.12

1. Check that the counter counts when a 10 Hz signal from the astable module is applied.

2. Connect a flying lead to the hold input. Connect the free end of this lead to the low voltage level. What happens?

3. Disconnect this flying lead from the low level. What happens?

4. The reset input on this module is normally low, and has to be taken high to reset the counters. (This is the same as on the 4-bit binary counter you met earlier.) Try resetting the counter with the hold input high. What happens? Now try with the hold input low. What happens?

If the reset input is low (or not used) and the hold input is high, the counter counts, but the display is held when the hold input is taken low. During the time it is low, the counters *continue to count* the input. When the hold input is taken high again, the display continues from the count reached, not from the latched value. If the reset input is connected to the high voltage level, the counters are reset to zero, but, of course, the display only changes to 00 if the hold input is high.

Timing

In the applications at the end of this chapter, the dual decade counter module is usually used with the astable unit on the pulser/astable module. In fact the astable and dual decade counter were designed as a pair of modules which could together be used for a wide range of timing and display purposes. By using different astable outputs connected to the clock input of the dual decade counter (see Fig.19.12), the following time intervals can be displayed:

between 0 and 99 s   in steps of 1 s   using 1 Hz input signal
between 0 and 9.9 s   in steps of 0.1 s   using 10 Hz input signal
between 0 and 0.99 s   in steps of 0.01 s   using 100 Hz input signal
between 0 and 99 ms   in steps of 1 ms   using 1 kHz input signal.

Applications using counting circuits

Project 19(a) Divide-by-*N* counting
Build a divide-by-256 counter from two 4-bit binary counters. Use the counter to measure the frequency of the a.c. mains.

Project 19(b) Down-counting (1)
Using a 4-bit binary counter, design a circuit which counts down from 15 (binary 1111) to zero (0000). Use LED indicators to display the count.

Project 19(c) Down-counting (2)
Using a 4-bit binary counter, design a circuit which counts down from 7 (binary 111) to zero (000) and then returns to 7. Display the count with a seven-segment indicator.

Project 19(d) Event timing
Using an LDR, a NAND gate, an astable and a dual decade counter measure the velocity of a dynamics trolley.

Project 19(e) Interval timing
Measure the acceleration due to gravity by using a timing system built from an astable, a dual decade counter, an RS bistable, NAND gates and two LDRs. The system should start counting when a light beam falling on the first LDR is interrupted, and should stop counting when a beam falling on a second LDR is interrupted. When the system is working properly, mount the two LDRs vertically, one above the other, and measure the acceleration due to gravity by dropping an object to interrupt the two beams.

Project 19(f) Counting the 'swings' of a pendulum

Build a circuit which will count the 'swings' (that is, $\frac{1}{2}$-cycles) of a pendulum. Use a 4-bit counter and a seven-segment decoder/display module for this.

Project 19(g) (*very hard*) Counting and timing a swinging pendulum

Build a system which will automatically count five 'swings' (that is, $\frac{1}{2}$-cycles) of a pendulum and display the time taken.

Project 19(h) Controlling a motor (1)

Build a circuit which turns an electric motor on for 10 s in every 20 s.

Project 19(i) Controlling a motor (2)

Build a circuit which turns an electric motor on for 5 s in every 20 s.

Project 19(j) Reversing a motor at regular intervals

Design a control circuit for an automatic liquid mixer. The circuit should reverse the direction of rotation of the mixer motor every 5 s.

Project 19(k) Flashing a lamp six times – six 'pips'

Design a circuit which will flash a lamp six times when a switch is pressed and released, using the 1 Hz output from an astable. Then modify the circuit so that it will sound a buzzer six times.

Project 19(l) An automatic light-buoy

A warning buoy in a shipping channel is to have a light which flashes. It is to be on for 1 s and off for 4 s. Design a circuit which will do this. Use the astable/pulser module to provide 1 Hz pulses. Then adapt the circuit so that the light only operates when it is dark.

Project 19(m) An electronic die

Build an electronic die using clocked RS bistables, an astable, a seven-segment decoder/display, NAND gates and a latch. The display should normally cycle from 1 to 6 at high speed. When the hold input on the latch is taken low, the display should freeze and show the result of the 'throw'.

Project 19(n) Traffic lights

Build a control circuit for a set of traffic lights. Use the 1 Hz astable output for all timing, and the red, yellow and green LEDs on the indicator module for lights. The lights must come on in the normal traffic light sequence (yellow, red, red and yellow, green, and back to yellow again).

Project 19(o) (*very hard*) Batch counting

Articles in a factory are delivered from an assembly room to a packing room along a conveyor belt. Design a system which will stop the conveyor belt (turn off its motor) every time a batch of five articles has entered the packing room. Provide a switch in this room which can be used to restart the conveyor belt motor once the articles have been removed from the belt for packing. A second counter should count the number of batches of five articles sent to the packing room during the day.

Project 19(p) Reaction time

Build a circuit which can be used to test people's reaction times. A switch should be provided which starts a dual decade counter counting 100 Hz pulses from an astable. This switch is operated secretly by the person controlling the test. A second switch should freeze the display. This is operated by the person under test directly the display is seen to change. The dual decade module should then show the time difference between the two switch closures, that is, the reaction time of the person under test.

Project 19(q) (*very hard*) Frequency meter

Build a meter which will measure frequencies from 0 Hz to 99 Hz. This can be done by holding a NAND gate open for one second at a time by taking one of its inputs high, and applying pulses from the source of unknown frequency to the other input. A dual decade counter can be used to count and display the number of pulses occurring during the 1 s the gate is open. This number (the number of pulses per second) is equal to the frequency of the source.

In practice, the display must be frozen at regular intervals so that readings can be taken. Try to design a system in which the hold and follow times of the latch on the dual decade counter are both equal to 1 s. The display will then be updated every other second.

Project 19(r) Digital-to-analogue conversion

Fig.19.13 is a circuit to investigate rather than a problem to solve.

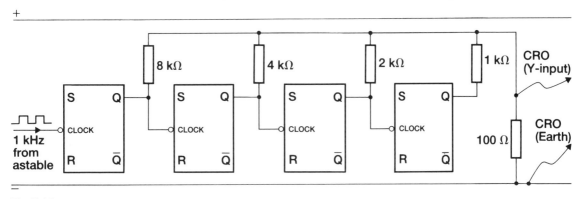

Fig.19.13

Build the circuit and then complete the investigations and questions below.

1. Connect a 1 kHz signal from the astable module to the first clock input, and observe the voltage waveform across the 100 ohm resistor on an oscilloscope. Sketch the waveform.

2. Try to explain why the waveform has this shape.

3. Suggest a use for a voltage waveform of this shape.

Chapter 20 **Memory**

Electronic memories are very important in modern electronics. An essential part of a microcomputer, for example, is its memory. When you load a program from a cassette tape or a disk into a microcomputer, it is stored in this memory.

An important aim of the present course is to develop some understanding of electronic memory and its use. In the last chapter we investigated a very simple memory – the hold/follow latch. But this could be used to store only one 4-bit binary pattern at a time. The electronic memory module (Fig.20.1) which we will be using in the present chapter can store up to sixteen 4-bit binary patterns at a time. It is much more like the memory used in a modern microcomputer, except that even a small microcomputer's memory is likely to be able to store thousands, or perhaps hundreds of thousands of binary patterns!

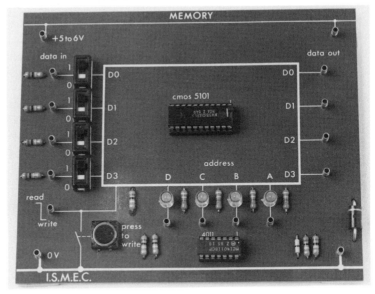

Fig.20.1

Before we start our experiments with the memory module, we need to introduce some new terms.

The electronic memory integrated circuit on the memory module can be thought of as a series of storage boxes stacked one on top of the other. Each storage box is called a memory *location* and each location has its own *address* which distinguishes it from the other locations. The 4-bit binary pattern stored in the location, the *contents* of the location, is known as *data*. Of course, this data may represent information in the same way that your binary patterns did in Experiment 16.1.

Address of the location D C B A		Typical contents of the location
0 0 0 0	contains	1 0 1 1
0 0 0 1		0 0 1 0
0 0 1 0		1 1 1 0
⋮ ⋮ ⋮ ⋮		⋮ ⋮ ⋮ ⋮
1 1 1 1		0 1 0 0

The memory module

In the case of the particular integrated circuit on the memory module, there are sixteen memory locations, each with its own distinct address. The address of the first location is the binary number 0000, the next the number 0001 and so on, as shown above, up to 1111.

In this module, every location can store a 4-bit pattern as data. In the next experiment we shall see how to put such a pattern into a memory location. We shall also see that, once a pattern is in a location, it remains there (it is memorised) until something is inserted in its place.

Experiment 20.1   Using a memory integrated circuit

The purpose of this experiment is to show how data can be stored in a memory, and how it can be retrieved at a later time. The first step is to set up the circuit in Fig.20.2.

1. Make the RESET input to the binary counter high. This resets the binary counter outputs A, B, C and D to zero, so all the LED indicators should be off. The counter outputs are connected to the address input lines of the memory, so the address on these lines will be 0000. When a location is addressed, the contents of the location appear on the four 'data out' lines and will be shown on the 'data out' display. Of course, in the present

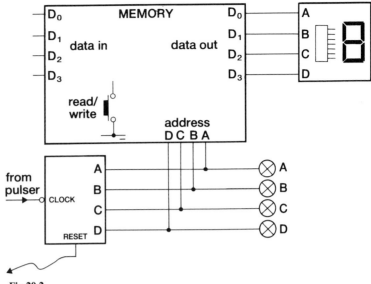

Fig.20.2

experiment, we are using a seven-segment decoder/display to show the contents of the addressed location and that means our display is limited to the binary numbers 0000 to 1001 (0 to 9 in scale-of-ten).

2. Press the pulser once. The address indicators should read 0001, and the contents of location 0001 will be displayed at the 'data out' display. In this way the contents of all the sixteen memory locations can be examined. This is called 'stepping' through the memory.

3. The next stage is to see how data can be inserted into a memory location. There is a 'read/write' push-button switch on the module. Normally the read/write line is high. In that position, data can only be read from the memory location. But if the push-button is pressed, the read/write line is taken low, and in that position whatever is present on the 'data in' lines will be stored in the memory location in place of what was there before.

4. Try storing your date of birth in the memory. Suppose, for example, this is 26th August 1972. The number to store would be 260872.

First you should address location 0000. Then set the data input lines D_3 to D_0 to 0010 to represent the number 2. The voltage levels for the data input lines can be selected by means of the miniature switches on the module, or by flying leads connected to the input sockets (switches set to the 'high' position). In the latter case note that data input lines float high, so a connection is only required for a low input. Take the read/write line low by pressing the push-button switch on the module. This stores the data, which is then shown on the seven-segment display.

5. Now use the pulser to increase the address by one so that location 0001 is looked at. Store the next number, in this case 6, by setting the data input lines to 0110. Then press the read/write push-button again.

6. Continue in this way until all six numbers have been stored. Reset the counter to zero and check that the date of birth is stored correctly in the first six memory locations.

7. You found in Experiment 16.3 that the seven-segment display is blank when all the inputs to the decoder are high. Now store the data so that, when stepping through the memory, the display shows, for example, ****26*08*72**** where * stands for 'display blank'.

Understanding memory in terms of simpler building blocks

The fabrication of ICs with many thousands of memory locations on a single chip of silicon is one of the most important developments in electronics and one of Man's greatest technological achievements. Now that you have met the logic gate, the decoder and the hold/follow latch, it is possible to show how a memory IC might be built up from simpler building blocks. You have seen all the blocks necessary for this, except one. The additional circuit is a tri-state gate, which can be thought of as an ordinary gate with a switch (Fig.20.3.).

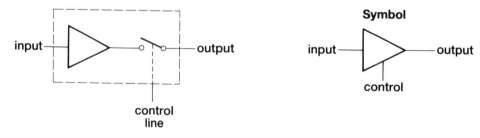

Fig.20.3

A control line operates the switch. When the switch is closed, the output will be high if the input is high, and low if the input is low. The device has the normal high/low output states.

But if the control line is used to open the switch, the output is not connected to anything. This is the third output state of the tri-state gate. In what follows we will assume that a high level on the control input disconnects the output, while, with a low level on this line, the output follows the input.

You may be worried that any disconnected line will behave as though it were high. Later you will see that there are many of these tri-state gates connected to the same output line, and there is always one (and only one) whose control switch is closed. It is that one which fixes the voltage level on the common line.

Having described how tri-state gates behave, we are now in a position to understand how a memory capable of storing sixteen 4-bit binary patterns might be built up. Consider the diagram Fig.20.4.

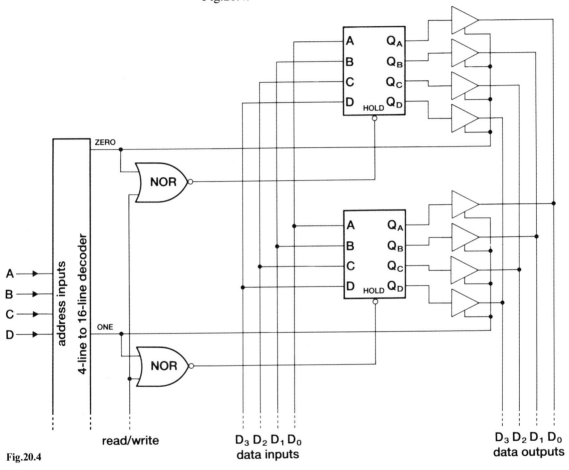

Fig.20.4

read/write

D₃ D₂ D₁ D₀
data inputs

D₃ D₂ D₁ D₀
data outputs

At the heart of the system there are sixteen hold/follow 4-bit data latches (for simplicity only two are shown here). On the left is a 4-line to 16-line address decoder. The four address lines can be used to select any one of sixteen output lines and therefore (as we shall see in a moment) any one of the sixteen memory locations. We will assume that the selected output line goes *low*, while all others remain high.

When address 0000 is present on the address lines, decoder output line zero goes low. Since this line is connected directly to the four tri-state gate control lines of the top latch, the four outputs of this latch are connected to the vertical data output lines. This vertical set of data output lines (known as a *data bus*) is connected to the outputs of *all* the tri-state gates at the outputs of all sixteen latches.

However, the outputs of the other fifteen tri-state gates will be in their third, unconnected state since their control lines are all high. Hence, although all the latches share the same data bus, only one (the one selected by the decoder) will be connected at any one given time.

Consider now the operation of the NOR gates. A truth table for any one of these will help us understand their action (Fig.20.5).

Input 1 (from read/write)	Input 2 (from decoder)	Output
1	1	0 (Hold)
1	0	0 (Hold)
0	1	0 (Hold)
0	0	1 (Follow)

Fig.20.5

The output of a NOR gate is high when neither input 1 NOR input 2 is high. From the truth table, we see that as long as read/write is high (the read state), the output will always be low. All data latches will therefore be in the hold state (we are assuming a low state for hold as for the latch module used earlier). If read/write goes low, we see that the output of a NOR gate with its other input low will go high. Only the NOR gate with its second input connected to the decoder output line selected by the address inputs will satisfy this condition. The latch connected to this particular NOR gate will therefore enter its follow mode, and will follow any data on the data input lines. All the other fifteen latches will remain in their hold state. So, by making read/write low we can store a 4-bit data pattern in the latch addressed by the decoder.

Types of memory

You have seen how an electronic memory can be built up from simple building blocks. The type of memory we have described is called a *random access memory*, or RAM for short. The importance of a RAM is that data can both be stored in it or retrieved from it at any time. Other types of memory known as *read only memories*, or ROMs for short, are widely used. The memory locations of these have fixed contents which cannot be changed. This is why they are *read only* devices.

Of course a ROM must once have been filled with data by someone, otherwise it would be useless! Usually this is done by the manufacturer of the piece of equipment using the ROM. For example, microcomputers often have the language BASIC stored in a ROM. It was put there by the manufacturer; you can use it, but you cannot change it!

When you load a program into a microcomputer from tape or disk, it is stored in the RAM of the computer. When the computer is turned off, the program is lost (on the other hand, the contents of a ROM are not lost when the power supply is turned off). A modern microcomputer will have thousands, or even hundreds of thousands of RAM locations. Each location will hold an 8-bit pattern, or in some of the more recent RAMs, perhaps even a 16-bit pattern. You may have heard of the term *byte*. A byte is simply an 8-bit binary pattern and is often used as a 'unit' of memory storage capacity. So, if a microcomputer has a 64 kilobyte RAM memory, this means it can store sixty-four thousand 8-bit binary patterns.

Applications using memory

Project 20(a) Traffic lights
Connect the red, yellow and green LEDs on the indicator module to the D_0, D_1 and D_2 data outputs on the memory module. Load the memory with the data necessary to light the indicators in the usual traffic light sequence, and then run the lights automatically using the 1 Hz output of the astable module linked to the clock input of a 4-bit binary counter connected to the address lines.

In the case of real traffic lights, red and green are on for a longer period than yellow or red/yellow. Build this feature into your system.

Project 20(b) Pedestrian crossing traffic lights
Build a circuit which will operate a set of pedestrian crossing traffic lights. These are normally at green. When a pedestrian wishes to cross, a push-button must be pressed. After a short delay, the lights go through the following cycle:

$$green \rightarrow yellow \rightarrow red \rightarrow flashing\ yellow \rightarrow green$$

Project 20(c) The BBC time signal
Build a circuit, using memory, which will produce the time signal when a switch is operated. The time signal consists of six 'pips', separated by equal intervals of silence, with the final 'pip' of twice the duration of the first five.

Project 20(d) Music from memory

Set up a circuit which can be used to store data in memory and then play back the stored binary patterns as a series of musical notes.

Project 20(e) An electronic die using memory

Build a circuit which will display, at 'random', a number between 1 and 6 when a switch is closed.

Project 20(f) Sine wave generator

Build a digital-to-analogue converter similar to that of Project 19(r) but using the data outputs of the memory module instead of the outputs of four clocked RS bistables. The circuit is shown below in Fig.20.6.

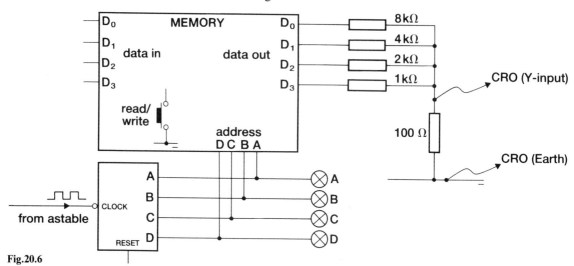

Fig.20.6

Your task is to load data into the memory which will result in the production of a sinusoidal voltage waveform across the 100 Ω resistor when the counter clock input is driven by one of the astable module outputs.

Project 20(g) (*very hard*) Automatic data logger

Build a data logging system which will automatically record the number of events occurring in regularly spaced 10 s intervals. The number of events occurring in the first 10 s interval should be recorded in memory location 0000, the number in the next 10 s interval in location 0001, and so on.

A system of this kind might be of use in a traffic survey of a remote road. The number of cars passing in regularly spaced intervals of an hour (or perhaps every hour) could be recorded over a period of several days. At a later date, the contents of the memory could be examined and used for analysis.

Background reading

Memories

Most of us buy memories quite often! They can be books, directories, records or tapes. They are stores for information which we call upon whenever we wish. The early computers used punched paper tapes as memories. Magnetic tapes and disks followed afterwards and 'floppy disks', which are smaller than rigid disks and made in flexible plastic material, are also being used.

Whereas our own domestic tapes and cassettes are magnetised according to the volume and character of whatever has been recorded, a computer's magnetic tapes or disks carry only a series of pulses. These correspond to the binary data they have received and which they will deliver when called upon to do so by the central processor. Each pulse is equivalent to a 'bit' which, in computing language, is the smallest unit of information. The capacity of a memory can, therefore, be measured by the number of bits it will hold.

Magnetic tapes and disks can store large numbers of bits but these tapes and disks have to be moved until the information required by the central processor is brought to the reading head. Consequently, it takes some time to extract information from this type of memory.

However, memories on chips are now in use which, as well as being contained in much smaller spaces, overcome time delays. Information is stored in different places on a chip and any one of these places can be reached in the same amount of time. These memories are called location-addressable and the time required to obtain information from them – the access time – is extremely small.

(From *The Challenge of the Chip*, HMSO.)

Chapter 21 **Why use digital electronics?**

What is digital electronics?

Most of the electronic circuits which you have met in this book have a common feature. They need some sort of input which is changed by the circuit to produce an output (Fig.21.1). The one exception to that is the astable circuit which produces an output without the need for an input. The changes, of course, are brought about by electrical methods and hence the name 'electronics'.

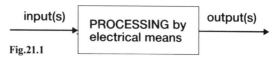

input(s) → | PROCESSING by electrical means | → output(s)

Fig.21.1

The input data has been encoded into binary form, processed and then decoded in some way to display the output or to use to drive motors etc. The information is processed only in the forms of 'low' or 'high', and there is no meaning to be given to anything in between. There are just the two voltage levels which the circuits can recognise and which have been linked to the digits 0 and 1, and so we call this kind of electronics 'digital electronics'.

Is there another kind of electronics?

The digital method is not the only way in which information can be processed. Suppose you wish to make the sound from a radio louder. You feed in the 'information' by adjusting the volume control and the output gets steadily and continuously louder. Most of the changes in the world around us are continuous changes: the change of light intensity from night to day, the rise in temperature of the water in a pan as it is heated, the transmission of a message by telephone; these are all continuous changes. And some electronic circuits process the 'information' in its continuous form.

To do that, the information has to be converted into a corresponding value of an electrical quantity such as voltage, which is then called the *analogue* of the input ('analogue' means 'corresponding to'). Of course the voltage has to be able to vary continuously at all stages of the processing and hence this kind of electronics is called 'analogue electronics'.

A microphone amplifier is an example of an analogue circuit.

When sound waves fall on the microphone, the microphone produces small voltages which change continuously according to the sound waves. Those changes then have to be enlarged in order to drive the loudspeaker. The circuit designer aims for a circuit which will produce an exact but magnified copy of the input variations. The difficulty is the exact part – no amplifier or loudspeaker does it *exactly* at all frequencies so that there is always some distortion or lack of accuracy.

What advantage has one method compared with the other?

We have already mentioned that analogue amplifiers suffer from lack of accuracy, even though the error may be small. This is true of all analogue circuits, as the following very simple example may show.

Suppose we wish to send information from one place to another along wires and that we have to transmit the number 109. We could use the circuit of Fig.21.2 and adjust the rheostat so that the first milliammeter read 10.9 mA. If everything was perfect, the more distant meter would also read 10.9 mA, but in practice the meters would not be exactly the same and someone has to read the pointer on the scale. It would be surprising if the information received were precisely 109.

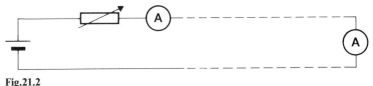

Fig.21.2

The same information could be sent in binary code (1101101). To do that would require eight wires between the two places as Fig.21.3 shows. That is a more complicated circuit than the one of Fig.21.2, but the information it transmits is precise since we only have to decide if the lamps are on or off. And if we want greater precision by transmitting more digits, we need only provide more switches, wires and lamps.

Fig.21.3

If electronic circuits are used in analogue systems, inaccuracies increase, but in digital systems that is not so. Compared with analogue circuits, digital circuits offer greater precision, though they are usually more complicated.

What about designing systems? The work you have done has shown how a simple building brick, the NAND gate, could be used to make up more complex building blocks such as the bistable, the 4-bit binary counter and the hold/follow latch. These larger building blocks were then used in the projects to construct systems for some definite purposes. This is illustrated in Fig.21.4. Designing a system is much easier in digital electronics, and not only because there are the larger building blocks available. Because we are only concerned with high and low voltage levels, there are no problems in connecting the blocks of a family together as there often are when one analogue circuit is connected to another. Furthermore analogue circuits need a larger number of external components.

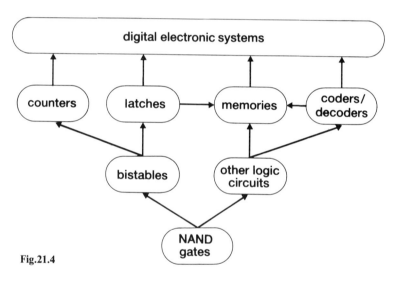

Fig.21.4

There is one snag with digital electronics however. The input information may come from the analogue world and so it has to be encoded digitally. Circuits which do this are called analogue-to-digital converters. And perhaps a digital-to-analogue converter will be needed at the output (see Projects 19(r) and 20(f) as examples). Fortunately both of these can be obtained in integrated circuit form, but it does increase the complication of the system.

Many systems require information to be stored and you have seen how the memory stores information in digital form. Analogue information can also be stored by using the voltage which is to be remembered to charge a component known as a capacitor. Unfortunately charge leakage causes the voltage across the capacitor to change and the stored information therefore changes and is

166

ultimately lost. A digital system has to be used if information is to be stored with precision.

The main advantage of digital electronics is the fact that the circuits can be prefabricated cheaply in micro-miniature form in an integrated circuit. Even complete systems can be treated in this way so that only the input and output connections have to be made. As mentioned above, this makes the designing of systems using ICs comparatively easy. You may wonder what it is about analogue circuits which make them more difficult to produce in IC form. Generally they need more external components than digital circuits and the values of these components have to be selected for the particular job in hand. It is therefore not possible to use a basic circuit and to build bigger blocks from it.

To summarise, the uses of digital electronics are growing rapidly because information processing is precise, the information is easily stored and the circuits can be prefabricated in IC form so making systems easier to design and small in size. The electronic clock described on page 140 was a system which involved eight 4-bit binary counters, two or three bistables, six seven-segment decoder/drivers, one 3-line to 8-line decoder and at least nineteen NAND gates. Even more would be needed for a digital wrist watch since a supply of 50 Hz pulses from the mains is not available. The wrist watch would need an astable to supply pulses and more dividing circuits. Yet all these circuits, the display and the setting switches can be put in a very small case to go on your wrist – and leave enough room for a cell to provide the power. That illustrates the advantages of IC digital electronics.

Where do we go from here?

We hope you have been impressed by the number and variety of the systems which can be constructed using the NAND gate as the building brick – and the ICs (which incorporate NAND gates) as larger building blocks. It is not difficult to believe that it lies in the power of men and women to invent a system to do anything they wished and then create it in IC form. A dramatic step forward was made when it became possible to produce an IC, which could be programmed, in other words instructions could be given to it. How might that be done?

To take a simple example, suppose it would be useful to have a circuit which could be any one of the four basic logic gates. Naturally, we should have to tell the circuit how we wanted it to behave. We could do that by having two extra inputs (the inputs X and Y in Fig.21.5) and fixing the voltage levels applied to X and Y according to what we wanted the circuit to do. Two inputs would be necessary because there are four different logic gates, and there are four different patterns of voltage level that can be applied to

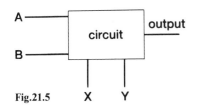

Fig.21.5

two inputs. The instructions to the circuit might be given as follows.

Level at X	Level at Y	Circuit behaves as
0	0	NAND gate
0	1	AND gate
1	0	OR gate
1	1	NOR gate

It is not a difficult job to build this circuit. It can be made from thirteen NAND gates and Appendix C (on page 180) shows the circuit and explains how it works.

Clearly, for the many tasks that our universal IC would have to do, a large number of instruction inputs (such as X and Y above) would be required as well as a large number of inputs (such as A and B) and many outputs too. Our universal IC would need to be like an electronic brain, capable of adding, subtracting, multiplying, dividing, classifying data, drawing conclusions, remembering words, translating languages, and so on, and all by supplying it with voltage patterns as an instruction. The number of NAND gates needed would be enormous!

Nevertheless, human ingenuity has devised a system which, with the help of a relatively small number of auxiliary circuits, will do many of the things mentioned above. This system is the micro-computer, with which you are no doubt familiar. Memories, counters, latches, displays, logic gates, all have a part to play in that system, but at the heart of the microcomputer is an IC which can be given instructions so that the computer will perform the many different functions required. This IC is called a *micro-processor*. Electronics designers are developing more versatile microprocessors, which will allow more information to be handled more rapidly, so that electronics will have an increasing effect on the lives of us all.

Background reading

▲ user port

Fig.21.6

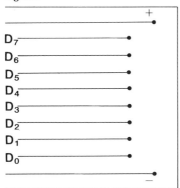

Fig.21.7

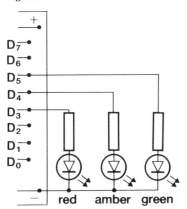

red   amber   green

Fig.21.8

Using a computer for control

It is also possible to use computers for control.

If you look carefully at most microcomputers, you will find that there is a socket, perhaps at the end, on one of the sides, or even underneath, called a 'user port' (Fig.21.6).

A port is somewhere where wires or conducting strips from the inside of the computer are brought to the outside. Most people working with computers use the word 'lines' to describe these conductors, so we shall do the same.

The number of lines brought to the user port is not the same for all computers, but in most it will be at least ten. Ten lines are shown in the diagram in Fig.21.7. Two of the lines are the positive and negative supply lines from inside the computer (like the positive and negative supply rails in your electronics). The other lines (labelled D_0 to D_7) can be made high or low by the computer in the same way that the output of a NAND module can be either high or low. A series of instructions can be typed into a computer to tell it which lines to make high and which low. This series of instructions is called a 'program'.

Suppose you wish to control a set of traffic lights. You would first connect the three lights to three of the lines, as shown in Fig.21.8. A series of instructions must then be given to the computer to make the lights come on in the right order. The instructions must do something like the following:

1. Make line D_3 high, the others low.

2. Pause one minute.

3. Make lines D_3 and D_4 high, the others low.

4. Pause ten seconds.

5. Make line D_5 high, the others low.

6. Pause one minute.

7. Make line D_4 high, the others low.

8. Pause ten seconds.

9. Go back to instruction **1**.

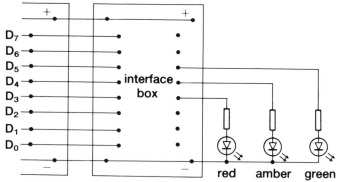

Fig.21.9

When this program has been put into the computer, the instruction RUN is given and the program instructions will be carried out one by one in the order given. The lights will therefore come on and off automatically, and they will keep on doing so, since instruction **9** makes the computer go back to instruction **1** to keep repeating what happens.

Of course, with a real computer you cannot just type in the words shown above. You must use the programming language for the computer, which is described in the manual provided with the computer. You will find there are words in the programming language which make the lines go high or low.

It is not a good idea to connect LEDs directly to the lines D_0 to D_7, as it is possible to damage the computer if wrong connections are made.

It is usual to have a special 'interface box' between the computer and the outside world. This is because this interface box usually contains special electronic circuits to protect the computer.

Other devices such as relays and motors can be connected to an interface box, and they can then be switched on and off using the computer lines.

You will have noticed that once again it is a matter of *switching* when working with computers like this, just as you were concerned all the time with switching when working with your electronics modules.

Chapter 22 **Some final questions**

1. A student sets up the circuit in Fig.22.1 as a decoding circuit. For what combination(s) of the input voltage levels at A and B will the output be high?

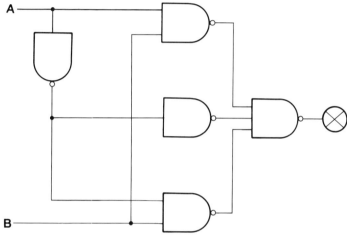

Fig.22.1

Devise a simpler circuit using only two NAND gates which will behave in the same way.

2. A simple encoding circuit is to have two outputs, X and Y, and three inputs labelled 1, 2 and 3. It is to operate so that the voltage levels at X and Y give the binary equivalent of the number of an input when that input is taken low. Design a circuit to do this.

3. A seven-segment indicator can be operated to display the first six letters of the alphabet (Fig.22.2).

Fig.22.2

Consider the problem of displaying the first four letters by means of the switches, correspondingly labelled A, B, C, D in Fig.22.3. Design a decoding circuit to do this using NAND gates. You may assume a segment is lit when the decoding circuit output is high. (The most economical solution requires three 2-input NAND gates.)

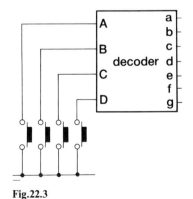

Fig.22.3

What does your indicator display if no switch is pressed?

Is it possible to display E or F with your circuit by pressing more than one switch? Explain your answer.

4. Describe how the circuit in Fig.22.4 works (a) if P is pressed and released with switch T in the open position, (b) if P is pressed and released with switch T in the closed position.

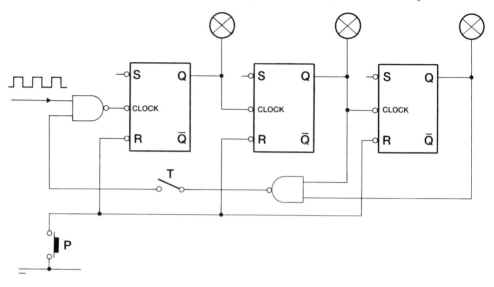

Fig.22.4

5. Draw a circuit diagram of a 4-bit binary down-counter using clocked RS bistables. Explain, with the help of a timing diagram similar to Fig.17.9 on page 131, how the circuit works.

6. Fig.22.5 shows a disc, made of insulating material, which can be rotated about its centre. Five spring contacts press against the disc, as shown. The shaded areas of the disc's surface are covered with a layer of copper, so that, as the disc is rotated, one or more of the contacts A, B, C, D, are connected to contact X. What might this be used for? Describe an application for this kind of switch in electronics.

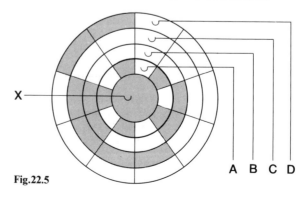

Fig.22.5

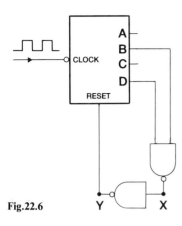

Fig.22.6

7. Fig.22.6 shows a BCD counter made from a 4-bit binary counter.

   a Draw up a table showing the voltage levels at each of the outputs A, B, C, D, after each of ten input pulses. Assume each output level is low to begin with.

   b What does 'BCD counter' stand for and why is it so called?

   c Why is it sometimes called a 'divide-by-ten' circuit?

   d 'The clocking pulse for another counter can be taken from the D-output, or from point X, or from point Y.' Is this statement correct? Is there any advantage in choosing point X?

8. Fig.22.7 shows a circuit which a student claims to be a hold/follow latch. The student says: 'If the HOLD input is high, gate 1 is open, and the voltage level at Q is the same as the voltage level at A. If HOLD is low, gate 1 is closed, and changes in the voltage level at A do not change the level at Q.'

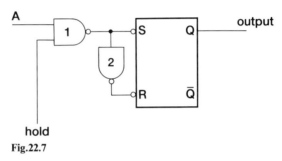

Fig.22.7

   a Show that the student's statement is correct.

   b Why is the circuit not a hold/follow latch?

9. A cheque is made for £24,369. Which is the most significant digit and which is the least significant digit?

10. 21603 is a number which might be displayed by five scale-of-N counters in sequence.

    a Which is the most significant and which the least significant digit?

    b What is the smallest value that N could have?

    c In a binary counter, each digit displayed is called a *bit*. Why?

    d What does a *byte* mean?

11. Fig.19.10 shows a block diagram of a dual decade counter. In each decade circuit, there is a hold/follow latch controlled by the hold input. It is suggested that these latches could be left out, and the BCD counters connected directly to the decoder/drivers, if the clock input was applied via a NAND gate as in Fig.22.8. How would the circuit then behave differently from that of Fig.19.10?

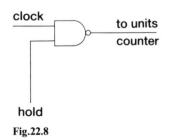

Fig.22.8

12. Fig.22.9 is the circuit of a frequency meter. Explain how it works. (*Hint*: draw a timing diagram to show how the voltage levels at Q, $\overline{Q}$ and the output of gate 3 change with time.)

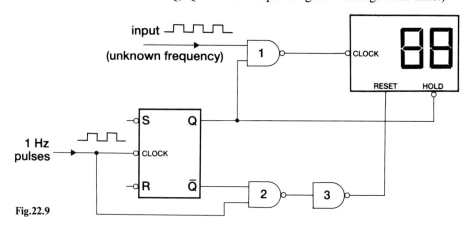

Fig.22.9

13. Describe what the circuits of Figs 22.10 and 22.11 do. In Fig.22.11, note that gates 1 and 2 are AND gates.

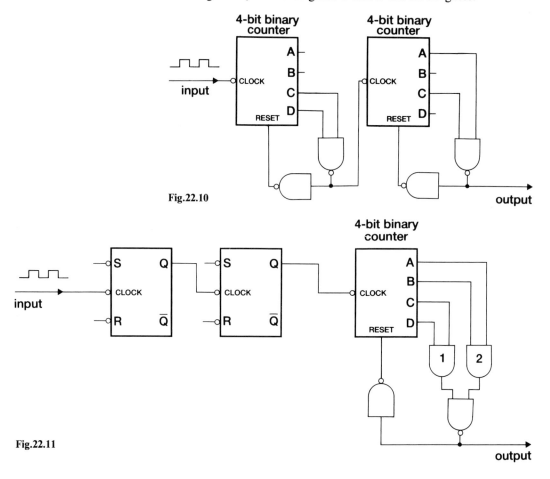

Fig.22.10

Fig.22.11

14. 'Electronics has fundamentally changed the lives of us all.'
a What social benefits has it brought?
b What economic benefits?

15. Write an account of the development of electronics. Your account might begin with the thermionic valve, refer to the transistor and continue to the integrated circuit and the microprocessor.

16. Has microelectronics contributed to the quality of our lives? To what extent may it, or may it not, do so in the future?

17. How might the microelectronics revolution affect the pattern of work in the future?

18. Describe the ways in which electronics has already contributed to life in the home and the ways in which it may contribute in the future.

19. 'Science in the nineteenth century was a concern of men. Microelectronics is as much a concern of women as it is of men.' Comment on this statement.

20. Is microelectronics a science? Is it engineering? Is it an art?

Chapter 23 **Appendices**

Appendix A
Switch contact bounce

In work with clocked bistables and counters frequent use was made of the pulser module to provide the necessary clock pulses. The purpose of the present experiment is to show why such a special unit is necessary, and the way it is constructed.

Experimental illustration of contact bounce

In the circuit of Fig.23.1, an ordinary SPDT switch is connected to the clock input of the 4-bit binary counter module.

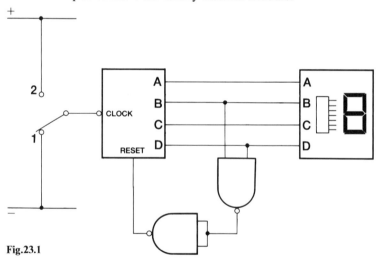

Fig.23.1

1. Set up the circuit and reset the counter to zero.

2. The moving contact of the SPDT switch is initially low. Push it from position 1 to position 2 and back to position 1 again. Do this several times and write down the number displayed after each operation. Does the count displayed increase in steps of one?

   In this experiment you will probably find that the counter will behave erratically – the count displayed may increase by more than one for each cycle of operation of the SPDT switch. This is because, like any ordinary switch, the SPDT switch is subject to *contact bounce*.

Contact bounce arises when the moving contact of the switch bounces on and off a fixed contact before settling in position. This results in the production of a series of pulses rather than a single 'clean' pulse (see Fig.23.2).

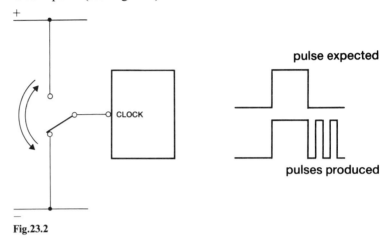

Fig.23.2

The extra pulses are generated when the switch goes to its low position, for a disconnected input behaves as though it is high. Thus, as the contact bounces, the input goes from low to high to low, etc. and real but unwanted pulses are generated. Contact bounce does, of course, occur when the switch goes to the high position, but in this case the bounce does not cause a change of voltage level at the input – unless the bouncing is so severe that the moving contact touches the low voltage contact again.

Contact bounce plays havoc with circuits such as clocked bistables and counters, because the extra pulses are real pulses which clock the system. Hence it is necessary to use some electronic circuitry to remove the effect.

Using a bistable circuit to eliminate contact bounce

Connect the SPDT switch to a bistable circuit made from two NAND gates, as shown in Fig.23.3. Note that the Q output of the bistable is connected to the clock input of the counter.

1. Use the SPDT switch to send pulses to the counter. Does the count displayed now increase by one each time?

2. Try to explain the action of the bistable in the circuit in Fig.23.3

In the present experiment two NAND gates connected as a bistable are used. From earlier work it is known that the R and S inputs are normally high, and that the R input must be taken

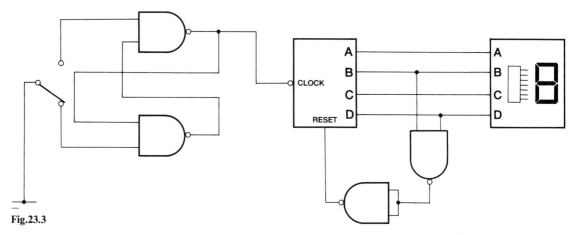

Fig.23.3

briefly low to enter the state with Q=0 and $\overline{Q}$=1. To enter the other stable state (Q=1 and $\overline{Q}$=0) the S input must be taken briefly low. The key point in the present application is that once the S or R input has been taken low a first time, taking the same input low a second or further time has no effect.

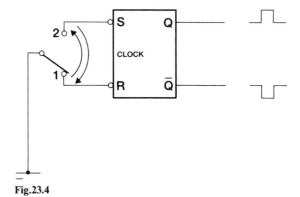

Fig.23.4

Thus, in Fig.23.4 if the bistable is initially in the state with Q=0 and the moving switch contact is pushed from position 1 to position 2, the S input will be taken low and Q will be set high. If the moving contact bounces on and off the fixed contact at 2, but not so severely as to touch 1 again, the S input will be taken repeatedly low, but this will not affect the state of the bistable (Q remains high).

When the moving contact is pushed back to position 1, the R input will be taken low and Q will be reset to zero. Again, contact bounce will have no effect because the bistable was reset at the moment of first contact and taking the R input low several times in succession produces no further change. In this way a *single* low–high–low pulse is produced at the Q output every time the moving contact is moved from position 1 to 2 and back to 1 again.

The single pulse output on the astable/pulser module used in this course is, in fact, made in this way to avoid contact bounce.

Appendix B
How to switch a digital clock from a 50 Hz to a 60 Hz input

If the input frequency is 50 Hz, pulses at the rate of 1 pulse per second are obtained from a divide-by-10 and a divide-by-5 counter in sequence.

If the input frequency is 60 Hz, the second stage needs to be a divide-by-6 counter in order to obtain 1 pulse per second.

Thus there needs to be a 'select frequency' pin, X, such that the voltage level changes a divide-by-5 counter (which is reset when $A=1$, $B=0$, $C=1$) into a divide-by-6 counter (which is reset when $A=0$, $B=1$, $C=1$).

Thus the counter must reset if $A=1$, $B=0$, $C=1$, $X=1$ or if $A=0$, $B=1$, $C=1$, $X=0$. A 3-input NAND gate will recognise $A=1$, $C=1$, $X=1$ by giving a low output only then. Note that a C input is necessary for otherwise the pattern $A=1$, $X=1$ with $C=0$ would cause reset.

Similarly $B=1$, $C=1$, $\overline{X}=1$ can be recognised and so give a low output from a NAND gate. If either of these NAND gate outputs are low, the counter must receive a high at its reset input. Another 2-input NAND gate will produce this.

Fig.23.5 shows the circuit. The output pulse for the seconds counter can be obtained from the C output, which goes low when the counter resets.

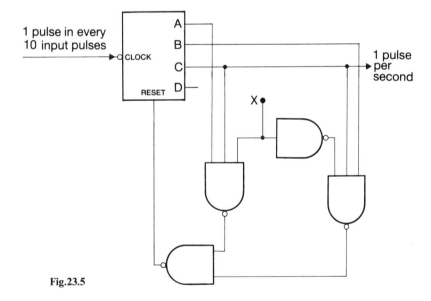

1 pulse in every
10 input pulses

CLOCK

RESET

A
B
C
D

X

1 pulse
per
second

Fig.23.5

Appendix C
A programmable logic gate

A circuit which will behave as described on page 168 is shown in the diagram below (Fig.23.6).

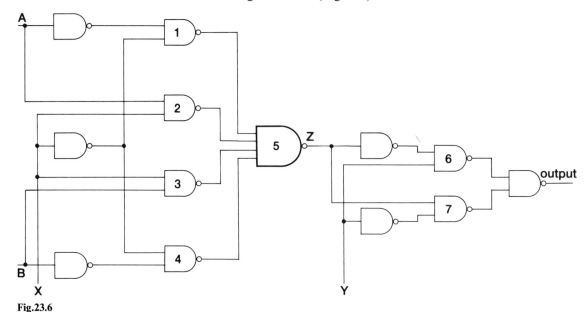

Fig.23.6

It is not difficult to understand how it works. If X is low, gates 2 and 3 are closed and their outputs are high, whilst gates 1 and 4 are open with their outputs the same as the voltage levels at A and B respectively. The inputs to gate 5 are therefore 'A, high, high, B' so that the output of gate 5, Z gives the NAND function of inputs A and B.

If X is high, gates 1 and 4 are closed (outputs high) and gates 2 and 3 are open. The inputs to gate 5 are now 'high, $\overline{A}$, $\overline{B}$, high' so that Z gives the OR function of inputs A and B (see page 79).

The Y input controls the part of the circuit to the right of Z. If Y is low, gate 6 is closed (output high) and gate 7 is open. The final output is therefore the same as Z. If Y is high, gate 7 is closed with output high and gate 6 is open. The output is then the inverse of Z, namely $\overline{Z}$. Thus:

if X = 0 and Y = 0,
    Z is the NAND function and the final output is NAND;
if X = 0 and Y = 1,
    Z is the NAND function and the final output is AND;
if X = 1 and Y = 0,
    Z is the OR function and the final output is OR;
if X = 1 and Y = 1,
    Z is the OR function and the final output is NOR.

Index

Ammeter, use of 2
Analogue electronics 164
AND circuit 18, 70
ASCII code 124
Astable 126

Binary code 114, 130
Binary coded decimal (BCD) 119, 138
Binary counter 130
Binary down-counter 132
Binary number 116
Bistable 52, 84
Bit 114
Burglar alarm 54
Buzzer 12
Byte 161

Cells in parallel 38
Chips 58, 61, 75
Circuit diagrams 7, 55, 68, 88
Circuits, combination of 78
Clocked bistable 128
Coding 114
Common anode display 118
Common cathode display 118
Computer for control 169
Control system 101
Counting circuits 136

Data 146, 156
Decoder 119
Decoding 116
Digital clock 140, 179
Digital electronics 164
Dimmer 5
Diode 6
Divide-by-N counter 132
Drivers 91
Dual decade counter 148

Electrical pressure 36
Encoding 116

Gating effect 68, 134

Hold/follow latch 145

Integrated circuit (IC) 58, 61, 63
Inverter 42, 48

Latch 145
Least significant bit (LSB) 116
LED indicators 66
Light dependent resistor (LDR) 11
Light emitting diode (LED) 9
Logic circuits 76

Measuring current 2
Measuring voltage 37
Memory 155
Microprocessor 168
Most significant bit (MSB) 116
Motor 13
Music box 123

NAND integrated circuit 63, 66, 76
NAND relay module 41, 47
NOR circuit 70, 77
NOT circuit 69

OR circuit 18, 70, 77

Power supplies 40
Problem solving 95
Programmable logic gate 180
Pulser 126

Quad NAND 63

Random Access Memory (RAM) 160
Read Only Memory (ROM) 160
Reed relay 21
Reed switch 20
Relay circuit 41
Resistance 5, 6
Rheostat 5

Seven-segment displays 116
SPDT switch 19
Speed of motor 26
SPST switch 19
Supply rails 11

Switch contact bounce 176
Switches 15
Symbols 7
Systems 101

Tea-maker 101, 106
Thermistor 72
Timing 151
Transistor 56

Tri-state gate 158
Truth tables 45, 67, 76, 79, 104

Unconnected inputs 68

Voltage divider 71
Voltage levels 39
Voltmeter, use of 38